GRACELAND CEMETERY

GRACELAND CEMETERY

A Design History

CHRISTOPHER VERNON

University of Massachusetts Press   Amherst and Boston

in association with

Library of American Landscape History   Amherst

LC 2011021546
ISBN 978-1-55849-926-3

Designed by Jonathan D. Lippincott
Set in Baskerville
Printed and bound by Thomson-Shore, Inc.

Library of Congress Cataloging-in-Publication Data

Vernon, Christopher, 1964–
  Graceland Cemetery : a design history / Christopher Vernon.
    p. cm.
  Includes bibliographical references and index.
  ISBN 978-1-55849-926-3 (cloth : alk. paper) 1. Graceland Cemetery (Chicago, Ill.)
2. Cemeteries—Illinois—Chicago—Designs and plans. 3. Chicago (Ill.)—Buildings,
structures, etc. I. Title.
  F548.612.G72V47 2011
  363.7'50977311—dc23

                                                                    2011021546

British Library Cataloguing in Publication data are available.

Photographs of Graceland Cemetery by Carol Betsch appearing on pages ii, vi, 14, 32, 48, 64,
86, 108, 132, 162, 167, 168, and 184, and in the gallery of color plates following page x, were
specially commissioned for this book. All photographs © Carol Betsch, 2011.

Contents

Preface

This book can be traced back to the efforts of three individuals who have long had an interest in Chicago's Graceland Cemetery putting its best foot forward to the public. Robert T. Isham Jr., John K. Notz Jr., and Frederick G. Wacker III, Trustees of the Graceland Cemetery Improvement Fund, all shared the belief that the cemetery would be well served by paying far greater attention to the history of its landscape than had been case for many years, at least since the 1920s. Although widely considered a pivotal design in the history of American landscape studies, Graceland had never before been the subject of a book-length study, and these individuals persisted in their belief that such a document would aid immeasurably in the pursuit of the cemetery's landscape goals.

A Walk through Graceland Cemetery, a guide written by Barbara Lanctot and published by the Chicago Architecture Foundation in 1977 (and revised several times since), was one of the first authoritative treatments of the site. In 1985, the eminent historian Walter Creese published an insightful essay on Graceland's design in his book *The Crowning of the American Landscape: Eight Great Spaces and Their Buildings*. However, neither of these publications delved into the layers of design that had been contributed by as many as seven landscape gardeners. In generally accepted lore, O. C. Simonds, who had the longest tenure at Graceland as superintendent, receives full credit for the design. When the Library of American Landscape History approached the trustees with a proposal to develop a definitive book on this influential and understudied landscape, the circumstances seemed ripe for such a project.

The trustees agreed with our recommendation that Christopher Vernon, an associate professor of Landscape Architecture at the University of Western Australia, be invited to write the book. Vernon had contributed a fine introduction to the LALH reprint of Wilhelm Miller's *Prairie Spirit in Landscape Gardening* (which features Graceland throughout) and he had also been a contributor to a 1992 report on Graceland Cemetery commissioned from the architectural firm of John Eifler.

The report by Eifler & Associates, in addition to laying the groundwork for the present study, set in motion other initiatives, one of which focused on the restoration of monuments and another on the restoration of the cemetery plantings "in the spirit of O. C. Simonds." That report also provided data that proved critical to the cemetery's 2001 nomination to the National Register of Historic Places, prepared by Legacy Historic Preservation Planning, the firm of Charles D. Kiefer, Rolf Achilles, and Neal A. Vogel. Other early scholarship was made available in *The AIA Guide to Chicago*, edited by Alice Sinkevitch (1993), with important contributions by Julia Sniderman Bachrach and Joan Pomaranc. This publication was the first to focus on the roles of the several landscape gardeners who had preceded Simonds at Graceland Cemetery: William Saunders, Swain Nelson, Horace W. S. Cleveland, and William Le Baron Jenney.

Over the course of Vernon's research, many new discoveries came to light. One of these was the extent of the real estate dealings of Graceland Cemetery Company, the original profit-making venture and the parent organization of the Trustees, a nonprofit organization. Vernon's account details the legal wrangling with the town of Lake View, which bitterly fought the cemetery's efforts to expand and also protested the holding of land destined for residential development, which exempted it from real estate tax. (This information was gleaned from the letters of Thomas Barbour Bryan, the founder of the cemetery, held by the Chicago History Museum, where many early plans of Graceland are also archived.) Vernon's text chronicles other dramas, as he analyses design developments including the construction of new roads, lakes, and buildings. He follows the cemetery's development under the leadership of Bryan's successor (and nephew) Bryan Lathrop, c. 1880–1898, when Simonds was using the landscape almost as a private laboratory to work out his distinctive planting style, which relied primarily on hardy, regional plants. Vernon suggests that Lathrop's influence began to wane even before the turn of the century and so did that of his protégé, Simonds, even though he continued to have an official role at the cemetery until his death in 1931.

An afterword by Ted Wolff, a Chicago-based landscape architect who has been advising on landscape improvements to the cemetery since 1990, brings the story of Graceland up to the present day. In 2009 Wolff was commissioned to create a long-term plan to provide ongoing guidance to development in the cemetery, again in the spirit of Simonds. The trustees have heartily endorsed the historical and horticultural principles embodied in this report, reaffirming their stewardship commitment to a landscape that not only provides a final resting place for many Chicagoans but also a historic, parklike sanctuary for the living.

I am very grateful to the Trustees of the Graceland Cemetery Improvement Fund for making this book possible, in particular Mssrs. Isham, Notz, and Wacker, who demonstrated great patience as LALH guided this project to completion. I thank Christopher Vernon, who rose to the challenge of completing this study with diligence and creativity. Vernon was assisted in the early stages of his manuscript by Joel Ray, a much valued developmental editor for LALH, and in the later stages by our project editor, Mary Bellino, who made important contributions to the final manuscript and also provided the index. The Chicago History Museum and the archives of the Chicago Park District furnished several images for the illustration program, which was skillfully coordinated by Jessica Dawson. I am, as always, grateful to our book designer, Jonathan Lippincott, whose patience and talent have benefited LALH for many years, and to Carol Betsch, whose evocative photographs of the cemetery add greatly to this book and also provide a record of the landscape for future historians. I am particularly indebted to Bruce Wilcox and the staff of the University of Massachusetts Press, publishers of this volume.

Robin Karson
Executive Director
Library of American Landscape History

GRACELAND
CEMETERY

Tulips, Graceland Cemetery.
Photography by Arthur G. Eldredge. Courtesy Chicago History Museum.

Introduction

Graceland Cemetery, laid out over several decades on a sandy ridge in northern Chicago, eventually became one of the best known landscapes in the world. In 1915, more than fifty years after its dedication, the parklike setting was identified by the horticulturist Wilhelm Miller as "perhaps the most famous example of landscape gardening designed by a western man." Miller rhapsodically continued: "It is more than a mere cemetery, for it is full of spiritual suggestion, and its wonderful effects produced by trees and shrubs native to Illinois have profoundly influenced the planting of home grounds."[1] Graceland's naturelike planting compositions also influenced the design of parks, campuses, and institutional grounds throughout the Midwest and beyond, bolstering an indigenous "Prairie School" of landscape design whose most famous practitioner was the Danish-born landscape architect Jens Jensen.

The planting compositions that transformed Graceland were the work of Ossian Cole Simonds, the cemetery's superintendent for more than three decades. Simonds used Graceland as his private design laboratory, experimenting with lyrical combinations of common trees and shrubs—such as oak, maple, ash, hornbeam, hawthorn, witch hazel, dogwood, and elder—many of which he transplanted from the wild. Carefully cultivated in the cemetery setting, these plants eventually achieved great stature and beauty, stirring not only aesthetic admiration but pride over the burgeoning American movement in landscape design. Photographs of Graceland were featured in many period publications, conveying an appreciation for the quiet beauties of native vegetation, preferred by "cultured" persons over the "showiest plants from all foreign lands."[2] In the years leading up to the First World War, nativist language laced increasingly polemical writings by Miller and others. Graceland's beauty was breathtaking, but more important to some, it had become a symbol of American purity.

To most visitors Graceland was also a source of solace and peace, a welcoming haven in an urban setting. The beauty of this place owed much to Simonds, but it also owed much to the layers of design that shaped the framework on which Simonds worked. This work was accomplished by a sequence of practitioners that included two looming figures in the history of American landscape, H. W. S. Cleveland and William Le Baron Jenney. These men, and those that came before, firmly believed in the salutary force of nature. And, in fact, legions of admiring visitors regularly made the trip to Graceland to lose themselves in the mirrorlike perfection of Lake Willowmere, to watch spring-blooming bulbs emerge under the wide-spreading branches of native oaks, and to observe native haws and maples take on the russet tones of autumn. They continue to do so today.

There is little question that Graceland's status as one of America's most important cemeteries is attributable to its conception as a work of landscape art. This book aims to recover the multilayered history of this iconic landscape. Design does not occur in a vacuum; along with the aesthetic vision of designer and client, the designs of gardens, parks, and cemeteries also encapsulate broader social concerns. Consequently, our study of Graceland Cemetery must begin with a consideration of its wider context.

Chicago's origins can be traced back to the late 1700s, with the founding of a trading post at the mouth of the Chicago River. In 1803 the U.S. Army built Fort Dearborn on a small hill overlooking the river. The site of the future city, as the architectural historian Robert Bruegmann describes it, was "an inhospitable stretch of marshy terrain that smelled of [wild] onions" on the southwestern shores of Lake Michigan.[3] The area was first used as a portage between the Great Lakes and the Mississippi River, which would enable the city's growth as a continental transportation hub. From its original lakeshore environs, Chicago—and soon the nation—would expand westward into one of

America's most distinctive natural landscapes, the prairies; although greatly altered by agriculture today, the gently undulating prairies once covered much of the midwestern and western United States and Canada. Chicago had just over four thousand residents when it was incorporated as a city in 1837—officially adopting the motto *Urbs in Horto*, or "City in a Garden."

Urbs in Horto registered more an aspiration than it did the fledgling city's reality. City boosters such as William B. Ogden (Chicago's first mayor and later a member of Graceland's founding board) saw the cultivation of gardens as a means to "demonstrate the city's cultural refinement," hoping that it would be seen as more than a business center.[4] By 1840 the city's population had swelled to almost 4,500; only a decade later nearly 30,000 people would call Chicago home.

In 1848 water transport routes expanded, and with them the city's role as a trading center. In that year the Illinois and Michigan Canal, linking Lake Michigan to the Mississippi, was completed; via river and canal, it was now possible to ship goods from the Great Lakes to the Gulf of Mexico. The same year saw the opening of Chicago's first railway, another conduit for manufactured goods and farm produce. Only the year before, as the historian Donald L. Miller notes, Chicago had not a single mile of track, but a decade later it was "the rail center of America."[5] As Bruegmann writes, "The city's development as one of the world's great railroad centers led to its dominance in the lumber, coal, and steel industries and the manufacture of farm implements, railroad cars, machinery, and household goods of all kinds."[6] Cattle operations followed the railroads' western expansion, and Chicago's vital role as a rail hub would soon lead to its becoming an epicenter for the meatpacking industry—Carl Sandburg's "hog butcher for the world." As its street gridiron quickly spread across the virtually flat terrain, the city also attracted people seeking lucrative business opportunities, especially in real estate.

View to Burnham Island.
Photograph by Arthur G. Eldredge. Courtesy Chicago History Museum.

One of those was a Virginia attorney, Thomas Barbour Bryan. Arriving in 1852 by way of Cincinnati, Bryan established a legal practice, but he soon came to specialize in land speculation and the buying and selling of real estate, often acting as an agent for other investors. One of his enterprises would be the founding of Graceland Cemetery.

Nearly four thousand Chicagoans died in the Civil War; some casualties would come home to their final rest at Graceland. But the war also became a catalyst for industrialization, facilitated by Chicago's maturing transportation links and commercial economy. As the historian Theodore Karamanski notes, Chicago's chief rivals, Cincinnati and St. Louis, were too near the front lines in the early stages of the war, and Chicago was able to eclipse them as a center for grain distribution, meatpacking, and the heavy industries that supported the Union Army.[7]

In 1871, with the Civil War and a martyred president still a recent memory, the now legendary Great Fire destroyed roughly a third of the city's center (including the office of Graceland Cemetery and the records held there). But Chicago would rise from the ashes like a phoenix; the disaster became a catalyst for rebirth and further growth, and Chicago boosters now set out to rebuild the city under the banner "I Will." Of this period Bruegmann writes: "Chicago was experiencing changes of all kinds faster than the more established cities in the East or in Europe. It was a tumultuous place, building and rebuilding itself so fast that at times it must have seemed the entire city was under construction. It was the 'Shock City' of the 1870s and 1880s, where travelers went to view the future."[8] Mark Twain, in *Life on the Mississippi* (1883), called Chicago "astonishing"; he described it as "a city where they are always rubbing the lamp, and fetching up the genii, and contriving and achieving new impossibilities." "It is hopeless," Twain observed, "for the occasional visitor to try to keep up with Chicago—she outgrows his prophecies faster than he can make them." The city's population had jumped to over half a million by the time Twain penned his description, and when Graceland's landscape development was completed around 1900, the figure had more than tripled.

"In Chicago," as the cemetery historian Helen Sclair astutely observes, "the living and the dead have always sought the same space, high and dry land with good transportation." In the early

Lake Willowmere.
Photograph by Arthur G.
Eldredge. Courtesy Chicago
History Museum.

days, she notes, "both populations shared settlements at Fort
Dearborn and along rivers." In 1835 the fledgling town established two burial grounds at its eastern edge. Soon enough, however, "the dead were standing in the way of the living," and both
graveyards were closed. By 1843 Chicago had established the new
City Cemetery on the lakeshore in what was then its northern
hinterland (now Lincoln Park).[9] Cholera epidemics in the 1850s
increased the cemetery's population and also fueled concern that
ongoing burials there would contaminate the city's water source,
Lake Michigan. It was against—and perhaps in entrepreneurial
response to—this backdrop that Chicago's first rural cemeteries
began to appear, further north of the city. Rosehill Cemetery saw
its first burial in July of 1859, and the Roman Catholic cemetery
Calvary was consecrated the same year. The dedication of Grace-

land would follow in 1860. By 1866 Chicago had abandoned City Cemetery, and bodies were exhumed for reinterment in rural cemeteries, including Graceland.

Unlike their urban predecessors, Chicago's new rural cemeteries were no mean burying grounds. By the nineteenth century, as the art historian Sally A. Kitt Chappell notes, most Americans had "changed their ideas about death since the Puritan preacher Jonathan Edwards spoke of 'sinners in the hands of an angry God.'" By contrast, "sermons—and paintings and poems—spoke of victory over death and the continuity of family life in a land of celestial bliss, but the belief had no expression in landscape art until the advent of the rural cemetery movement."[10] By the time it appeared on the Chicago scene, the rural cemetery idea was already decades old and had covered considerable distance in finding its way from its European birthplace to the American prairies. Chappell writes: "Early American rural cemeteries were inspired by Père-Lachaise, the immensely popular rural cemetery in Paris, created in 1804 in the English picturesque landscape style. In the newly industrialized nation, rural cemeteries like Mount Auburn in Cambridge, Massachusetts (1831), Green[-]Wood in Brooklyn, New York (1838), and Spring Grove in Cincinnati, Ohio (1845), provided some of the few accessible parklike spaces for people who lived in crowded cities. Weekend revelers flocked to their lawns for picnics, family get-togethers, and other summer activities."[11]

Graceland and Chicago's other rural cemeteries would soon prove no less popular. Indeed, as Donald Miller observes in his history of the city, these cemeteries had become so popular by the 1860s that "one of them began charging admission to families failing to produce a burial-lot certificate." Paradoxically, as Miller points out, "to get some peace and quiet, to breathe pure and wholesome air, the citizens of the city of the living had to visit the gardens of the dead."[12]

But the city was not as welcoming of art as some. As a character in a novel by the Chicago writer Henry Blake Fuller (1857–1929) put it: "This town of ours labors under one peculiar disadvantage: it is the only great city in the world to which all its citizens have come for the one common, avowed object of making money. There you have its genesis, its growth, its end and object. In this Garden City of ours, every man cultivates his own little bed and his neighbor his; but who looks after the paths between? Chicago, then, is not the sort of city in which artists are encouraged."[13] Indeed, whereas its Cambridge and Cincinnati predecessors had been founded by horticultural societies, Graceland would be established by private investors. Graceland and Chicago's other rural cemeteries did, however, provide an early civic amenity as quasi-parklands; it was nearly a decade after Graceland's founding before the city's park districts were established. In the end Graceland would become Thomas Bryan's object lesson that "making money" and the production of beauty were not mutually exclusive pursuits.

Graceland also displays, as Chappell writes, "a panorama of historic attitudes toward interment."[14] These attitudes, too, changed through time. Older portions of the cemetery, for instance, register the mid-nineteenth-century practice of delineating individual plots with iron railings or stone coping. This approach would later be supplanted by concern for creating an overall, harmonious composition of tree-framed lawns—"a soft green blanket over gently contoured land"—free from railings and coping, with low, inconspicuous gravestones.[15] Nonetheless, even this newer phenomenon did not remain static. Beginning around the turn of the century, a series of grandiose monuments were constructed to famous people by famous people, some of them seemingly oblivious of the cemetery's wider landscape setting.

The cemetery's landscape layout is best understood as a palimpsest. Perhaps even more so than its counterparts elsewhere, Graceland in its first half-century was a highly contingent, evolving creation involving no fewer than six primary designers—Swain Nelson, William Saunders, H. W. S. Cleveland, John A. Cole, William Le Baron Jenney, and O. C. Simonds—each man working over the work of his predecessors. Saunders made the first layout and Nelson implemented it; Cleveland and later Cole orchestrated expansions; Jenney laid out the cemetery's final expansion, draining and sculpting the terrain and providing Simonds with a foundation upon which to plant. Also important were the ideas and guidance of a pair of talented amateur landscape gardeners, Graceland's progenitor, Thomas Barbour Bryan, and his nephew—and successor as cemetery president—Bryan Lathrop. In fact, when Simonds began work at Graceland under Jenney's tutelage in 1878, he had no experience in landscape gardening. The canvas on which these designers worked was dynamic, as new land was acquired from time to time.

Graceland's midwestern location is in itself significant. Initially the prairie grasslands were dismissed aesthetically because of their flatness and visual monotony. Landscape gardeners, professional and amateur alike, who came to Chicago from Europe or the eastern United States usually recreated scenic vignettes more typical of Old World gardens or New England. Such was the case with Graceland's early layouts. Indeed, all but one of the cemetery's designers were "transplants" to the Midwest. And yet one must remember that in the mid-nineteenth century landscape gardening—or landscape architecture, as it is known today—was a fledgling profession with only a handful of practitioners. By the century's end, however, local architects such as Louis Sullivan had cultivated an aesthetic appreciation of the prairies. In his *Autobiography of an Idea* (1924), Sullivan (writing in the third person) recalled his first encounter with the prairies, which "utterly

Lake Hazelmere.
Graceland Cemetery (Chicago: Photographic Print Co., 1904), courtesy Trustees of the Graceland Cemetery Improvement Fund.

amazed and bewildered him" as he viewed them on a rail journey from Philadelphia to Chicago in the 1870s: "Stretching like a floor to the far horizon, —not a tree except by watercourse or on a solitary 'island.' It was amazing. Here was power—power greater than the mountains. Soon Louis caught glimpses of [Lake Michigan], . . . superbly beautiful in color, under a lucent sky. And overall spanned the dome of the sky, resting on the rim of the horizon far away on all sides, eternally calm overhead, holding an atmosphere pellucid and serene."[16]

Sullivan's appreciation of the prairie landscape would eventually spawn—driven by his protégé Frank Lloyd Wright—a regional school of design with Chicago as its epicenter. Landscape architects would follow. Paradoxically, one of the most eloquent articulations of this new regional landscape appreciation was written by a pair of Harvard academics. In their seminal textbook, *An Introduction to the Study of Landscape Design* (1917), Henry Vincent Hubbard and Theodora Kimball wrote:

> The sea alone, or a great lake, can vie with the prairie in the overwhelming simplicity of its effect. Extent, vastness, are alike in prairie and sea, but while the sea is always alive, even if at times asleep, the prairie is dead. It is immovable, ponderous, monotonous, stupefying. Each slight undulation which bounds the view gives promise of something different beyond, a promise always unfulfilled as one swell of ground succeeds another through days of travel. But nowhere better than on a prairie are to be seen the glories of the powers of the air. The squadrons of towering white cumulus clouds, giving in their diminishing perspective even a vaster sweep of view than the land, the daily miracles of sunset and sunrise, the clean and exhilarating summer breeze, or the deadly fury of a prairie blizzard, give to a man in an unusual degree a sense of standing directly in the presence of the great forces of the natural world.[17]

Within this shift in notions of landscape beauty or taste, Graceland would prove to be a harbinger of change. Graceland was at its landscape pinnacle by the second decade of the twentieth century, when Arthur G. Eldredge recorded it in a series of luminous photographs. Even if only unintentionally, the cemetery's undulating landscape of sinuous drives and water bodies, velvety greenswards, and groves of predominantly native trees and shrubs would soon become a touchstone for a new generation of landscape architects. In its final form, Graccland's design linked the nineteenth-century work of landscape gardening pioneers to a second generation of designers. With restoration efforts ongoing and a burgeoning interest in native plants and sustainable design, Graceland Cemetery has no less potential to inspire twenty-first-century landscape architects.

ONE

Thomas Barbour Bryan and the Genesis of Graceland

The idea of Graceland Cemetery originated with the Virginia-born attorney Thomas Barbour Bryan (1828–1906), whom contemporaries remembered as a "brisk, energetic little man, capable in affairs" and "widely erudite in language and literature."[1] (Fig. 1.1) Graduating from Harvard in 1848, Bryan wed Jane Byrd Page two years later, and he established a law practice in Cincinnati, where he remained for two years. Attracted by prospects of "a more lucrative practice" and "financial opportunities in real estate," Bryan moved to Chicago in 1852.[2] By the mid-nineteenth century, railroads had quickly put the fledgling city on the map, and it now was poised to become, perhaps even more than St. Louis, the "gateway to the West." (Fig. 1.2) The city Bryan encountered belied its motto, *Urbs in Horto;* far from being a garden, Chicago was, as one resident put it, "a mud hole from one end to the other."[3] From the beginning Bryan tried to help civilize this "crude young city," serving as president of the YMCA, building one of the city's important cultural centers, and even running for mayor twice.[4] He supplemented his legal activities with banking and real estate investment, and quickly achieved financial success and social prominence.[5]

Bryan fostered the city's emergent arts culture through his patronage of the celebrated portraitist George P. A. Healy (1813–1894). In 1860 he erected Bryan Hall, opposite the courthouse in the city's heart.[6] Fronted with a marble entry and adorned with frescoes, this domed auditorium had multiple functions, accommodating commercial offices (including Bryan's own), and seat-

1.1. Thomas Barbour Bryan (1828–1906).
Library of Congress.

1.2. Chicago in 1858. The Sherman House hotel, at the northwest corner of Clark and Randolph streets, was lost in the Great Fire but was soon rebuilt as one of the city's most luxurious hotels.
Photograph by Alexander Hessler, *McClure's Magazine,* December 1906.

ing over five hundred in its main hall, a capacity then unmatched in Chicago.[7] The building also served as an art gallery of sorts; Bryan lined the walls of its public spaces with patriotic works by Healy, including a collection of portraits of U.S. presidents.[8] He also maintained a private gallery in the building, opening it to the public on request. By the time Bryan Hall was lost to the Great Fire in 1871, it was renowned as one of the city's most prominent musical, theatrical, and oratorical centers.[9] When Ralph Waldo Emerson made a lecture tour in 1863, for instance, Bryan Hall was his Chicago venue.[10]

BRYAN AND LANDSCAPE GARDENING

Bryan's cultural initiatives were not confined to the city. He was no less concerned with improving the rural hinterland and eventually the suburbs the railroads generated in their wake. Four years after his arrival in Chicago he had accumulated enough wealth to purchase a one-thousand-acre tract in neighboring Du Page County.[11] Sixteen miles to the city's west, the locale was known as Cottage Hill—attractive partly owing to its elevation, some one hundred feet above the city, and its level environs.[12] At around the same time he also purchased the Grove, a farmstead east of Cottage Hill, just over the border in Cook County.[13] The motivation for these land acquisitions was not purely speculative, for commuting by rail enabled Bryan to retreat from urban life, and by 1857 he was in residence at the Grove. For reasons that remain unclear, however, he soon grew disenchanted with the farmstead, and the following spring he began developing a new pastoral retreat on his Cottage Hill property, naming it Bird's Nest (often mistakenly spelled Byrd's Nest)—an appellation derived from his wife's family name.[14] Perhaps seeking expert advice on this new enterprise, he entertained a remarkable houseguest that summer. In late June, Bryan reported the estate's progress to his brother-in-law Andrew Wylie in Virginia: "Nothing new since I wrote to you yesterday except the arrival last evening of Mr. Saunders, Landscape Gardener, who is to spend some days with us. He pronounces *Bird's Nest* where I have built a most charming site for a country residence, and prefers it [on account] of view and commanding position to the *Grove*."[15] William Saunders of Philadelphia was one of the very few professional landscape gardeners then practicing in America, and he would soon gain national prominence. Bryan would later hire him as one of the first designers of Graceland.

By 1859 Bryan's palatial new twenty-one-room home was finished. (Fig. 1.3) He soon filled it with Healy's paintings and other artworks, as well as tapestries and fine rugs gathered from around the globe.[16] Bryan could now pour his energies into domesticating the mansion's treeless surrounds, launching what one retrospective account described as "an elaborate system of landscape garden experimentation."[17] Eventually his grounds would include groves, rolling lawns, carriage drives and pathways, an orchard, and a "Sylvan Lake." (Fig. 1.4) There was, however, one marked departure from the English picturesque naturalism that otherwise characterized the place. The expansive flower garden was geometrically arrayed around an ornate white marble fountain acquired by the Bryans while honeymooning in Italy.[18] (Fig. 1.5) Bryan routinely opened his luxuriant gardens to his neighbors on Sunday afternoons.[19]

Owing to its prairie location, scale, and landscape effects, Bryan's estate attracted national attention in 1859. In March of that year, John J. Smith (1798–1881) met Bryan and visited Bird's Nest while on a visit to Chicago from Philadelphia.[20] A librarian and amateur horticulturalist, Smith was then the editor of *The Horti-*

1.3. Bird's Nest.
From *Horticulturist and Journal of Rural Art and Rural Taste* 15, no. 1 (January 1860).

culturist and Journal of Rural Art and Rural Taste, a position he held from 1855 to 1859. Started in 1846 by the pioneering landscape gardener A. J. Downing (1815–1852), the *Horticulturist* was the most important landscape gardening periodical of the day, and its editor was nationally esteemed as a horticultural authority.[21] Indeed, it is possible that Bryan first sought Smith's counsel via correspondence when he was looking for a landscape gardener for Bird's Nest the previous year. As Smith was one of William Saunders's most important patrons, it may well have been the landscape gardener who called Bird's Nest to his attention. Along with his literary pursuits, Smith had cofounded Philadelphia's Laurel Hill Cemetery in 1836 and still served as its president. For Bryan, Smith's achievement would prove portentous.

Bryan later lamented that Smith "could not have visited the prairie country at a more unpropitious season, for a succession of heavy rains had submerged the country."[22] And, far from being finished, the grounds of Bird's Nest were "torn up for tree-planting, road-making and the like."[23] Despite its incompleteness and "forbidding appearance," the Chicagoan's emergent landscape garden so impressed Smith that he promoted Bird's Nest in the pages of the *Horticulturist* some two months after his visit. In the May 1859 issue, Smith explained to his largely eastern audience that Bryan's efforts demonstrated that the prairie could be made a "paradise."[24] The level prairies of the Midwest, so unlike the comparatively well-wooded and rolling terrain of the eastern states, were then popularly regarded as posing an aesthetic liability to be overcome; no less a figure than Frederick Law Olmsted, for instance, would describe the country around Chicago as "not merely uninteresting, but during much of the year, positively dreary."[25] But Bird's Nest inspired Smith: "The poets have been educating us with the love of the heather, and have made the nightingale and whippoorwill household thoughts; who shall say that when the song about the prairie grass and the prairie birds has been as well and as long sung, we shall not admire and poetise them as much; there is plenty of heather land in Scotland not half so desirable as you will find on the level grounds of the West."[26]

Writing at Smith's request in the July 1859 issue of *Horticulturist*, Bryan outlined his own rather unusual appreciation of the region in the form of a letter to the editor, titled "Defence of the Prairies." He opens by pointing out that "mountains figure more pleasantly in the eye and on the canvas of the painter than profitably in the domain of the husbandman." Unlike the cultivation difficulties posed by rugged terrain, the flat prairies, Bryan writes, had "rich, deep loam all ready for the plough, and promising exhaustless fertility," and "the depth of soil favors the most luxuriant growth of grain, and grass, and flowers." As these agrarian concerns reveal, utility underpinned Bryan's aesthetic appreciation. "As I write," he notes, "an almost limitless expanse of prairie stretches out before me, bounding the vision only with the horizon. The undulations, the groves and creeks prevent monotony. . . . [T]his vast plain is clad in verdure of such depth of tone and with bloom of such exquisite hues, that I am confident were you here to witness it, and to see the occasional groves in richest foliage, and vast flocks of grazing sheep and cattle diversifying the scene, you would agree with me in pronouncing it enchantingly beautiful." He ends the letter on a more down-to-earth note: "But one of my men has come in to tell me that some thousands of feet of tile (for the lawn) and an underdrain mole machine (for general use in the farm) have arrived, and I must be out to look after them. I have been planting extensively, &c., &c."[27]

The gentleman landscape gardener's hurried conclusion is a telling one. Despite Smith's and now his own encomium to the prairies, Bryan's aesthetic cues were not indigenous. Instead, he was refashioning the local landscape in conformity with the genteel, sylvan imagery of the English landscape garden (as filtered through American sources such as Downing's writings) and employing "thousands of feet of tile" to replace the "exquisite hues" with expansive, monochromatic lawns. Trees and shrubs, not prairie flowers and grasses, were the plantings. At Bird's Nest, the prairie was important only as a background to be viewed from a distance. Nearly half a century would pass before Chicago spawned an indigenous, regional landscape design movement and the adaptation of eastern landscape taste gave way to innovation.

Bird's Nest also won local accolades. In July 1859, the same month Bryan's "Defence of the Prairies" appeared, a Chicago newspaper assessed his estate as especially "fine," noting that its owner "spends much of his time and a large amount annually in improvements." The account predicted that "in a few years, when his trees shall have more age," Bird's Nest would be "the most extensive private residence near Chicago."[28] In January 1860, in an article titled "A House in the West," the *Horticulturist* again extolled the virtues of Bird's Nest and even featured a full-page illustration of it.

1.5. The gardens at Bird's Nest around 1880. Courtesy Elmhurst Historical Museum.

1.6. Huntington's picturesque surrounds in 1873. Jedediah and Mariana Lathrop are seated on the veranda.
Courtesy Elmhurst Historical Museum.

In 1857 Bryan's metamorphosis at Cottage Hill had already attracted one of his friends to his bucolic haven, when George Healy purchased a portion of the property, including the cottage which gave the village its name.[29] Healy soon cultivated his own landscape garden at his new place, which he named Clover Lawn.[30] Bryan's brother-in-law Jedediah Hyde Lathrop (1806–1889) and his wife, Mariana Bryan Lathrop, followed in 1864, purchasing twenty-six acres from Bryan and similarly developing a picturesque estate, Huntington.[31] (Fig. 1.6) Later, beyond the confines of this growing enclave, Bryan, Lathrop, and other neighbors adorned the village's thoroughfares with American elms (*Ulmus americana*); Cottage Hill was renamed Elmhurst at Bryan's suggestion in 1869.[32] Thomas Bryan continued to subdivide his extensive holdings, and the rural village soon became a garden-filled suburb, with the lawyer esteemed as its "father."[33]

THE RURAL CEMETERY COMES TO CHICAGO

With Bird's Nest essentially complete by 1859 and Cottage Hill expanding into a village, Bryan's landscape gardening and real estate pursuits coalesced in a new enterprise, the creation of a rural cemetery. His decision to embark on this project was undoubtedly informed by his contact with John J. Smith. Given his preoccupations with rural improvement and landscape gardening, Bryan perhaps was a *Horticulturist* reader of long standing when its editor called on him on that wet day in March 1859. A visit to Bird's Nest, however, was not the reason Smith made the trip. He had actually been invited to Chicago to inspect the site of a new cemetery, Rosehill. And he did not journey from Philadelphia alone; Rosehill's landscape gardener, William Saunders, newly appointed on Smith's recommendation, accompanied him.[34] Although Smith did not record the presence of others, Saunders may well have joined him on his Cottage Hill excursion. A month later the *Horticulturist* trumpeted Saunders's prestigious new commission to lay out "the great cemetery of Chicago."[35] Indeed, in his May 1859 account of Bird's Nest, Smith reported that the "rural cemetery" Rosehill was then being established "on high land, once no doubt the shore of the lake, and with a soil and timber admirably adapted to its purposed occupation." "I predict," Smith concluded, "that this cemetery will become one of the greatest boasts of Chicago."[36] Although he may not yet have revealed it to Smith, Bryan had already located a tract for his own cemetery further south on the very same former lakeshore and would soon commission William Saunders to design it.

Taking its cue from the Parisian cemetery Père Lachaise (1804), a new type of burying ground emerged in 1830s America and soon began to supplant the tombstone-studded church and municipal graveyards of the colonial era. Increasing urbanization and consequent health concerns, such as cholera and smallpox, fueled the displacement of

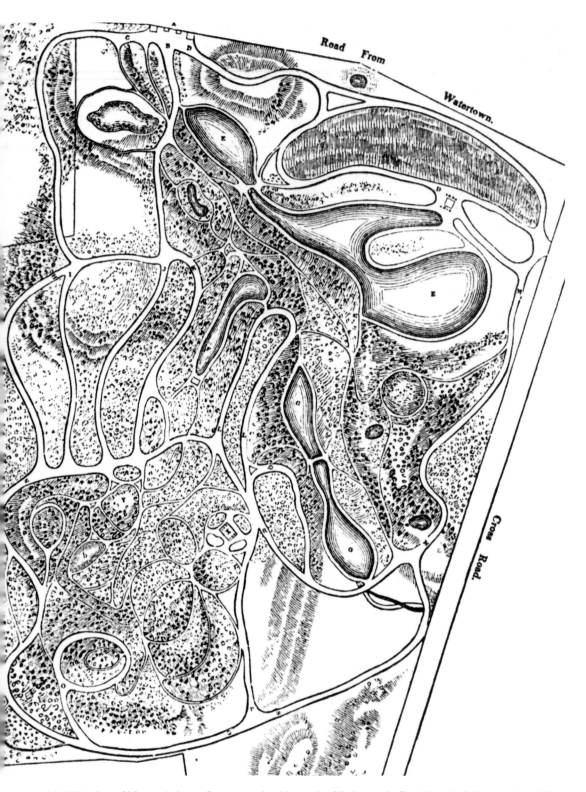

1.7. 1831 plan of Mount Auburn Cemetery, by Alexander Wadsworth. Pendleton's Lithography, 1831.
Courtesy Mount Auburn Cemetery.

city burial grounds to the rural outskirts. A shift in attitudes toward the dead was another potent factor in the proliferation of the new cemetery type. Beginning in the early nineteenth century, relative apathy—as registered by the often derelict, overcrowded state of urban graveyards—gave way to reverence. A belief that contact with some semblance of the natural world markedly improved the quality of urban life also arose at about the same time. These changes, coupled with the rising popularity of landscape gardening as influenced by English picturesque naturalism, coalesced to produce the rural cemetery.[37]

Even the term "rural cemetery" was novel in the 1830s. Derived from the Greek for "sleeping chamber," the word *cemetery*, Keith Eggener observes, "was rarely used by American-English speakers before this time."[38] The adjective *rural* was not simply a locational reference. "Contemporaries," Aaron Wunsch reminds us, "expected 'rural' in this context to denote a combination of 'genteel,' 'picturesque,' and 'suburban.'"[39] Mount Auburn in Cambridge, Massachusetts, laid out in 1831, was the first of the new rural cemeteries. (Fig. 1.7) Philadelphia responded with Laurel Hill in 1836, and Brooklyn with Green-Wood in 1838. (Fig. 1.8) All were "rolling, romantic environments, located several miles from their cities

1.8. Philadelphia's Laurel Hill Cemetery around 1848. Laurel Hill would become a design touchstone for Graceland. Courtesy Library Company of Philadelphia.

and managed by non-sectarian corporations."[40] Forerunner of the public park, the rural cemetery soon became a widespread phenomenon and began to move west. Cincinnati, for instance, gained one in 1844 with the establishment of Spring Grove. Three cemeteries would most inform Graceland's evolving landscape: Mount Auburn and Laurel Hill initially, and later Spring Grove.

The rural cemetery idea had found its way to Chicago by 1853, when the Oak Woods Cemetery Association was founded (although Oak Woods itself was not laid out until 1864).[41] (Fig. 1.9) The development was timely, as the city then desperately needed new sites for burial; cholera continued to take a heavy toll, for instance, with particularly severe outbreaks in 1849 and 1854. And the burgeoning metropolis had begun to encroach on City Cemetery, which had been laid out in 1842 well beyond the city's northern limits, on a lakeside site north of North Avenue and east of Clark Street, in an area within today's Lincoln Park. City Cemetery was Chicago's third municipal cemetery; in 1835 it had established a pair of burying grounds, one beyond the city at the north and another at its southern periphery. In 1843 the city closed both cemeteries to new burials and ordered that the existing graves be exhumed and relocated to the new City Cemetery. A few years after the removals were completed in 1847, none other than Thomas Bryan purchased the former southside cemetery and subdivided it for residential development.[42]

In 1858, with Chicago pushing at the edges of City Cemetery, local physician John H. Rauch spearheaded a movement to end burials within the corporate limits and relocate the cemetery.[43] For Rauch and others in the medical profession, the "emanations of the dead" were "injurious to health and destructive to life."[44] These gases, it was commonly held, facilitated the spread of diseases such as cholera. As trees were thought to absorb harmful "emanations," burials at a distance in leafy rural cemeteries were seen as a way of remedying these concerns. There was also another, more immediate health-endangering problem that mandated the closure of City Cemetery. Its low-lying position near the lake meant that it was cyclically inundated with high water, and it was not uncommon for coffins to resurface as a result of hydrostatic uplift pressure and the freeze-thaw cycle. The cemetery's proximity to the lake also raised concerns that further burials might contaminate the city's water supply. In the fall of 1858, Rauch recorded, "a number of

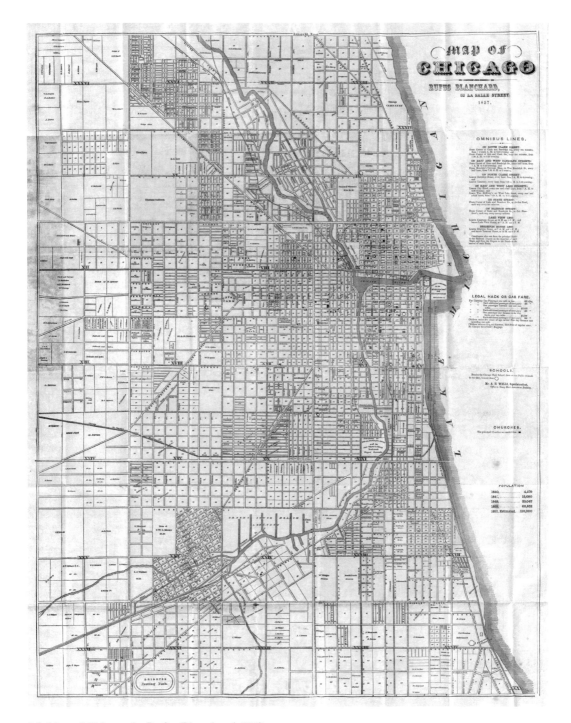

1.9. Map of Chicago by Rufus Blanchard, 1857.

Courtesy David Rumsey Map Collection, www.davidrumsey.com.

the most prominent and influential citizens of the North Division" petitioned the city's aldermen (called the Common Council), "remonstrating against further interment in the cemetery."[45] The council responded in November, appointing a special committee to consider "the removal of the present cemetery grounds to new locations."[46] There was now official impetus to establish new cemeteries, and the hunt for sites was on. Calls for relocation and the resulting official reconnaissances in 1856 and 1857 had not met with success, but collectively they suggest that Chicago's need for new burying grounds would have been widely known—especially to land speculators such as Thomas Bryan.[47]

The formation of the special committee "attracted the attention of several of the gentlemen corporators" of what would soon be Rosehill Cemetery and apparently prompted them to action.[48] And, despite the ethical ambiguities, public and private interests would quickly merge. The Rosehill Cemetery Company gained its charter on February 11, 1859, and four days later the committee recommended to the Common Council the "adoption of an ordinance directing the Mayor, Comptroller, and City Clerk, to confer with the managers of the Rosehill Cemetery with reference to the interment of those whom the city should be obliged to bury, and directing that the sale of burial lots in the City cemetery should cease."[49] Two weeks after the committee's report, as we know, John J. Smith and William Saunders were in Chicago to inspect the future cemetery's grounds. Eleven days later, as the fast-acting "gentlemen corporators" had apparently anticipated, the Common Council endorsed the special committee's recommendation.[50] Rosehill was inaugurated on July 28, 1859.[51] In November, the city gained another new burying ground with the opening of a Catholic rural cemetery, Calvary, in the northern suburb of Evanston.[52] Graceland would follow the next year.

When John J. Smith and William Saunders toured Rosehill in March 1859, Thomas Bryan had had an interest in rural cemeteries for nearly four years, owing to the death in April 1855 of his infant son, Daniel Page Bryan.[53] In need of a burial plot, Bryan had purchased one at the City Cemetery. Bryan later wrote, "The neglected and actually repulsive condition of the cemetery induced my search of land for a rural burying ground, more remote from and more worthy of the city." He formulated a preliminary plan for a new rural cemetery and began to gather support for it among his fellow citizens, but the plan was shelved indefinitely after Rosehill was established. Perhaps acknowledging Bryan's plan as preceding their own, the Rosehill Cemetery Company offered him its presidency.[54] For unspecified reasons, however, he declined. Thus Bryan was familiar with Rosehill's activities when he met their consultant John J. Smith the next month. If, as Bryan claimed, Rosehill's invitation "revived" his own project, then perhaps it was Smith who now convinced him that "citizens would be benefited by having two cemeteries between which to make selections."[55]

THE SITE

Although the specific purchase date remains unclear, we know that at some point between late 1858 and the beginning of 1860 Bryan acquired a 122-acre tract from Justin Butterfield and his heirs.[56] Unlike the comparatively level, treeless prairie around Cottage Hill, Graceland was to be developed on a series of well-wooded "rising ridges," north of the city and about a mile west of Lake Michigan in the town of Lake View.[57] Fronting Green Bay Road (now Clark Street), then a main artery, Bryan's tract lay at the northeast corner of the junction with Albert Street (now Irving Park Road). Likening it to Laurel Hill and Mount Auburn in the Graceland charter, Bryan touted the property as being "situated at a convenient distance, five miles from the Court House by carriage (over an excellent gravel road), and yet sufficiently remote from the City, two miles north of the City Limits, never to be encroached upon, no matter what may be the prosperity of Chicago."[58] Rosehill Cemetery was located one and a half miles to the northwest and serviced by the newly opened Chicago and Milwaukee Railroad.[59] And Calvary Cemetery lay just beyond, along the same rail line. Although Graceland was then accessible only by carriage, a railroad would later link it to the city.

The ridges Graceland would soon occupy are segments of a larger landform, geologically termed a spit—a narrow point of land projecting into water.[60] This undulating terrain was produced by the retreat of continental glaciers during the geological period known as the Pleistocene. Once a beach, the spit registers the fluctuations of glacial Lake Chicago (now Lake Michigan); at its maximum extent, about 12,500 years ago, this lake covered what is now the entire city of Chicago. The spit that includes Graceland was formed around 8,000 years ago, when the ancient lake was approximately twenty feet higher than the current lake.[61] The site's highest elevation is a ridge along its western or Green Bay Road boundary. The trajectory of this road, originally a Native American trail, followed the ridge line. From here at the west, the terrain slopes or rolls down to the property's eastern limits in a series of subtle undulations. Bryan's surveyors identified the landforms as ranging in height from "17 to 26 feet above the present stage of water in Lake Michigan." "Most of the ground," they noted, was "over 20 feet above," making it "four times the usual depth of graves."[62] Unlike the clay soils typical elsewhere in Chicago, the landforms here consist of almost pure sand. Ideal for a cemetery, this permeable composition affords good drainage, comparatively easy excavation, and, especially when supplemented with topsoil, a more hospitable medium for cultivating trees and shrubs. Farms in the vicinity were renowned for their production of celery, a vegetable that thrives in sandy soil.

Compared to the old City Cemetery's low-lying, marshy terrain, the topography of the future Graceland was dramatic. The site, Bryan later boasted in the cemetery's charter, was "admitted by all to be peculiarly beautiful," and the "gentle undulations," together with the "grove of old trees over the entire subdivision," suggested the name Graceland. The property was "high, diversified, and wooded," the charter stated, "broken into alternate though light elevations and depressions of unequal and irregular form, presenting every variety of surface that could be desired. The depressions are not, as is generally the case in this flat region, marshy basins or sloughs, but present dry, gravelly beds, of the same character as the surrounding heights. The whole is covered by a heavy growth of oaks, patriarchal in age and appearance."[63] Bryan had astutely selected a property in which utility and beauty converged, and, at least for one of the cemetery's later officers, the site's "natural beauty invite[d] the added graces which cultivated art bestows."[64]

Although Thomas Bryan believed the site to be "sufficiently remote" from Chicago itself, it did have one possibly unforeseen flaw. While prosperous farms occupied the western environs of the property, to the east Bryan's tract abutted land already in the midst of suburbanization. In 1854, the developers James Rees and Elisha Hundley selected this locale for transformation into a "lakeshore community of fine homes and landscaped grounds," and they erected what they hoped would be the future suburb's nucleus, Lake View House. This hotel, or perhaps more accurately lakeside resort, was calculated to attract potential investors in local land.[65] It succeeded, for in 1857 original settlers—immigrants primarily from Germany, Sweden, and Luxembourg—and new suburban arrivals organized the vicinity into the Township of Lake View. After opening in 1860, Graceland became "a destination for outings" from Lake View and would help attract further development there.[66] By 1865 its population would swell enough that Lake View would be incorporated as a town, and in 1887 it became a city; ultimately, in 1889, it would be annexed by Chicago. As Graceland and Lake View grew, so would conflict between the two.

TWO

The First Designers: Swain Nelson and William Saunders

Reflecting on Graceland's origins, Thomas Bryan noted that after identifying a site, he next began investigating cemetery design and "obtained from abroad the best works which had been published on the subject."[1] Though he identified neither the titles of the texts he consulted nor how long he spent studying them, we do know which works were available in the mid-1850s. The most prominent and directly relevant was John Claudius Loudon's *On the Laying Out, Planting, and Managing of Cemeteries and on the Improvement of Churchyards* (1843). The Scottish landscape gardener, as his biographer Melanie Simo writes, "believed that cemeteries should be sited just outside of towns, so that—properly laid out, 'ornamented' with tombs, and planted with labelled trees, shrubs, and herbaceous plants—the cemetery might become a 'school of instruction in architecture, sculpture, landscape gardening, arboriculture, botany and the important points of general gardening: neatness, order, and high keeping.'"[2] Presuming that individual graves would be marked and enclosed, he tended toward rectilinear configurations when making his own cemetery layouts. American cemeteries such as Mount Auburn, Laurel Hill, and later Graceland may have shared Loudon's broad ideals, but not his general designs. Although Bryan mentioned only foreign texts, by 1855 there was also a considerable domestic literature on rural cemeteries. John J. Smith had already published his own treatise on the subject, *Designs for Monuments and Mural Tablets, Adapted to Rural Cemeteries, Church Yards, Churches and Chapels* (1846). The first of its type in America, Smith's pattern-book was—by its author's own admission—derived from Loudon's treatise; as Aaron Wunsch notes, apart from excising some of Loudon's passages, Smith's contribution consisted of adding the occasional sentence and making substitutions such as "Philadelphia" for "London."[3]

Bryan may also have consulted a timelier American volume. In April 1855, the same month Bryan lost his son and began his cemetery study, the *Chicago Tribune* reviewed G. M. Kern's *Practical Landscape Gardening, with reference to the Improvement of Rural Residences, giving the General Principles of the Art; with full directions for Planting Shade Trees, Shrubbery and Flowers, and Laying Out Grounds* (1855). Praising it as "one of the best books of the kind ever published," the reviewer noted that it covered the "adaptation" of landscape gardening to cemeteries.[4] In a ten-page chapter on the subject, Kern advocated that cemeteries be "laid out in accordance with the most approved rules of the Art of Landscape Gardening"—although he did not explicitly state just what those rules were.[5] He also illustrated his text with a prototypical plan. (Fig. 2.1) Believing "light, fantastic shapes, or fanciful designs" to be "entirely out of place," Kern illustrated a layout that featured a fluid, elliptical network of carriage roads. From these roads, Kern wrote, "grass-walks or sweeping gravel-walks" should lead to individual lots, "giving free, unrestrained access to all." With respect to vegetation, "powerful masses of trees should be formed into shady groves, adding variety and beauty to the scene, [bringing] into bolder relief the smooth lawns devoted to graves." Plantings were to be composed with "a view to the prospects attainable, both to within and without, from different points on the grounds," and "conspicuous objects in the scenery should receive additional effect from the manner in which they are brought to the eye." Seeking to lend a degree of mystery to the grounds and illusively enlarge them, Kern advocated "partial disclosures," maintaining that "fine monuments or other objects" should not be allowed to "appear to the sight at once." He also urged that the practice of enclosing graves with "high, conspicuous railings or fences, iron or otherwise" be prohibited, as these were "entirely out of keeping with natural scenery."[6] Although picturesque compositions are comfortably familiar today, Kern's advocacy of planting trees in "powerful masses" marked a significant departure from convention. By contrast, authorities such as Loudon promulgated a planting design approach known as the gardenesque, in which

Fig. 15.

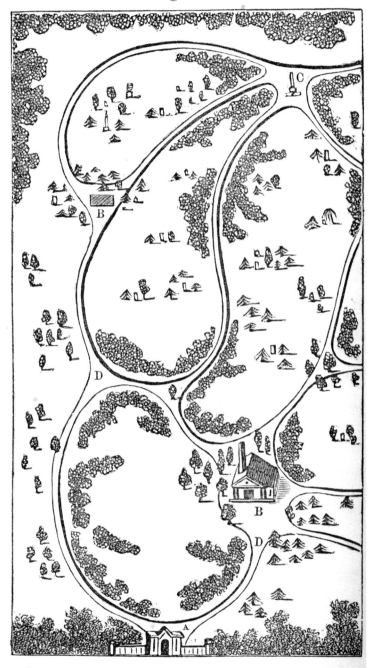

A, Main Entrance Portal.
B, Receiving-vault.
C, Public Monuments.

D. Main Carriage-way.
E. Exit Carriage-road.

2.1. G. M. Kern's prototypical cemetery layout, from his *Practical Landscape Gardening* (1855).

botanical specimens were isolated, enabling viewers to appreciate their individual horticultural beauty.[7]

Beyond the pages of *Practical Landscape Gardening*, little is known of G. M. Kern or his landscape gardening practice.[8] That the book was published in Cincinnati suggests the possibility of his presence there, yet although Kern cites Mount Auburn, Green-Wood, and an unnamed Kentucky cemetery, he makes no mention of Spring Grove or its designer, Adolph Strauch.[9] Still, his cemetery design ideals—especially his reference to "smooth lawns devoted to graves"—resonate strongly with Strauch's.

Around this time the *Horticulturist* also featured instructive essays on village and rural cemetery design. A. D. Gridley's June 1855 article, "Rural Cemeteries," is typical of these.[10] "Before a single stone is turned," Gridley wrote, those planning to develop a rural cemetery should secure a "professional landscape gardener"—that is, "an artist . . . who can appreciate all the capabilities of the place and can use them to the highest advantage." He should "prepare a plan suited alike to the nature and situation of the place, . . . and then his plan should be faithfully carried out." Gridley warned against "the folly of mapping off the ground into squares, like a checker-board, with straight roads and walks, and these bounded by stiff Balsam Firs at regular distances." Instead he advocated "main roads, leading by easy curves to all parts of the cemetery, and from these, gravelled walks [that] lead to every grave." The roads, Gridley writes, "will wind, not for the sake of winding, but because nature will indicate, here and there, that they should do so; as, for example, to avoid a tree, or hill, or rock, or pool of water."[11] John J. Smith also revisited cemetery design in the *Horticulturist*, contributing a three-part article titled "Rural Cemeteries" in 1856 and another with the same title in 1858.

By the time Smith's articles were published, Thomas Bryan had equipped himself and was ready to commission a designer. Possibly before he had even begun to search for one, Graceland's first landscape gardener would appear unsolicited on his doorstep.

SWAIN NELSON

In 1856, the year Thomas Bryan began Bird's Nest, the Swedish landscape gardener Swain Nelson (1828–1917) arrived in Chicago. Largely self-taught, Nelson was one of the very first to practice there,

2.2. James B.
Waller's home in
Lake View.
From Chamberlin, *Chicago
and Its Suburbs.*

and he would enjoy a lengthy career not only as a landscape gardener but also as a nurseryman.[12] According to his unpublished autobiography, one of Nelson's earliest Chicago projects was a residential design for James B. Waller (1817–1887), who, like Bryan, was a lawyer and real estate developer.[13] Relocating from Kentucky to Chicago in 1858, Waller purchased a twenty-acre tract in Lake View, and Nelson, noticing signs of house construction at the site, approached him and solicited the commission to lay out the grounds.[14] Along with planting and locating drives and paths, the project required Nelson to position the house, a barn, and a greenhouse on the developer's undulating property. (Fig. 2.2) Waller was evidently satisfied with the result as he continued his association with Nelson after the project was completed.

As Nelson relates the tale in his autobiography, at some point Waller told him that Thomas Bryan had purchased land adjoining his own and "intended to lay out a cemetery on it." Convinced that a cemetery would diminish his property's value, Waller confided, "I don't like it [and] I wish you could find out." Nelson approached Bryan the next day and inquired as to his plans for the cemetery— presumably without identifying the source of his knowledge. Bryan "was much astonished," as "he had not bought the land yet." That he had yet to acquire the land was not the only reason this stranger's questioning elicited surprise. As the *Chicago Tribune* would later reveal, Bryan had been attempting "purposely" to allow as little as possible about the cemetery to "transpire through the public press."[15] As the project was foremost a business enterprise, Bryan hoped to shield his activities from competitors such as Rosehill. The possibility of protest from those owning land in close proximity to Graceland, such as Waller, also mandated secrecy. Nelson did not record whether he conveyed the details of this meeting back to his patron. But Waller's resistance to Bryan's cemetery enterprise would prove ongoing.

On the strength of his successful design for Waller, Nelson solicited Bryan for work on the cemetery. "In a short time," Nelson recollected, "Mr. Bryan sent a letter for me to call at his office, when I met him in the office he said he had now bought the land and he wanted me to me[e]t him at his office next morning and we would go out together and look over the land." The two set out on horseback and made a thorough survey of the parcel. Nelson reported that Bryan had already secured a plan that located the cemetery's "entrance and office" and projected a "straight [entry] road [terminating at] a given point." Their joint reconnaissance culminated with Bryan instructing Nelson to extend and configure this entry road. Throughout the next two or three years, Nelson not only elaborated this drive but also laid out a second thoroughfare, grubbed out impeding oaks, and graded the roadbeds. When Bryan eventually requested yet another road, Nelson explained that he "could not work any longer without a[n overall] plan." "Well, make a plan," Bryan replied. Now effectively Graceland's first landscape gardener, Nelson "started to work out the whole ground in fifty foot squares" and developed a general layout, which Bryan approved. Nelson's association with the cemetery would continue for more than a decade.

As one would expect of a retrospective account made near the end of his life, Nelson's autobiography lacks chronological precision. He did not, for instance, identify specific dates for his Graceland activities. Nonetheless, he offers clues that help us begin to recover the exact sequence of the cemetery's development. Most historical accounts date Bryan's Lake View purchase to 1860. Nelson's autobiography, however, points to an earlier date. He writes that at the time he first met Waller, "people anticipated war if any change in the administration was made and it looked like Mr. Lincoln would be elected." And just above that passage he makes a clear reference to the Panic of 1857: "The times began to be hard. Banks were br[e]aking. The money we used was only paper money, and when we got a dollar bill you would not know if it was worth 100 cents or only 50 cents." And after describing his initial work at Graceland, he notes, "This together with other work gave me occupation between 2 & 3 years in the hardest times on account of the expectation of war." These indications would shift the date of Bryan's purchase two to three years before the outbreak of the Civil War in April 1861, that is, to 1858 or 1859.

Nelson's autobiography touches on other aspects of Graceland's origins. According to Nelson, Bryan's purchase was an individual, not a corporate, acquisition, and one made with the express purpose of establishing a cemetery. If Nelson's memory of the chronology of events was accurate, then months and possibly years passed between the land purchase and the point when Bryan was in a position to form the Graceland Cemetery Company. Corroborating Nelson's recollection, the cemetery company's charter records that its "title to the ground" was conveyed "from Thomas B. Bryan" alone.[16] Perhaps Bryan retained Nelson in advance of the company's formation because he believed some preliminary improvement was necessary to attract investment. Such a motive might also explain why he had secured a plan for Graceland's office and entry road layout before engaging Nelson.

WILLIAM SAUNDERS

Around 1858 or 1859, Graceland gained a second landscape gardener when Thomas Bryan commissioned William Saunders (1822–1900), "the eminent Landscape Gardener, of Philadelphia."[17] Born in St. Andrews, Scotland, Saunders "came of a family of noted gardeners, and throughout his early life he enjoyed exceptional opportunities to study botany, horticulture, and landscape gardening in his native country."[18] After horticultural study at St. Andrews and Edinburgh College, he "moved to London and worked as an apprentice gardener on several large estates."[19] Immigrating to America in 1848, Saunders soon began contributing articles to popular magazines such as the *Horticulturist*. In 1854 he established a partnership with the nurseryman Thomas Meehan (1826–1901) in the Germantown section of Philadelphia.[20] Saunders soon won national acclaim for his designs of country places, such as Johns Hopkins's Baltimore estate, Clifton

(1852), "regarded at that time as the finest private property in America," and public parklands, most notably Philadelphia's Hunting Park (1857).[21] (Fig. 2.3) By the 1860s Saunders had apparently begun to attract special esteem as a cemetery authority, designing, for instance, two cemeteries in New Jersey, Hazelwood Cemetery (1859) in Rahway and Alpine Cemetery (1862) in Perth Amboy.[22] (Saunders came to be credited with planning Philadelphia's Fairmount Park and Laurel Hill Cemetery, though he actually designed neither.)[23] The pinnacle of his cemetery design career was his layout of Soldiers' National Cemetery at Gettysburg (1863). In 1862, shortly after Saunders would complete his Graceland work, the U.S. Department of Agriculture appointed him to the post of Botanist and Superintendent of Horticulture. In this capacity he orchestrated the ongoing development of the national

2.3. William Saunders's plan of Hunting Park in Philadelphia (1857). Courtesy Fairmount Park Historic Resource Archive.

(*Opposite*) **2.4. William Saunders's plan for the campus of Illinois State Normal School (1857). Saunders would later produce similar layouts for Rosehill and Graceland cemeteries.** Courtesy Milner Library, Illinois State University, Normal, Ill.

capital's park system, designed a decade earlier by his acquaintance A. J. Downing.

Around the time Bryan commissioned him, Saunders had secured a constellation of Illinois commissions. Along with Rosehill Cemetery (1859) and Oak Ridge in Springfield (1859), he laid out an unidentified cemetery in Evanston (1859) and "embellished rural residences" for N. H. Ridgely in Springfield and Dr. J. A. Kennicott in Chicago.[24] Even earlier, in 1856 or 1857, Saunders designed Greenwood, the estate of Illinois businessman Jesse Fell, in North Bloomington (later renamed Normal). Fell facilitated Saunders's commission to lay out the newly established Illinois State Normal University (now Illinois State University), of which Fell was a founder (1857). (Fig. 2.4) Up until Rosehill Cemetery, this fifty-six-acre campus was Saunders's most extensive Illinois project. And in its similar network of winding drives and walks set amidst picturesque groves, the campus layout is not unlike his cemetery plans.[25]

Swain Nelson's autobiography mentions neither Saunders nor any other Graceland designer. If, as Nelson claimed, he was already at work on the cemetery's layout around 1858 or 1859, why did Bryan choose to involve another landscape gardener? At this junc-

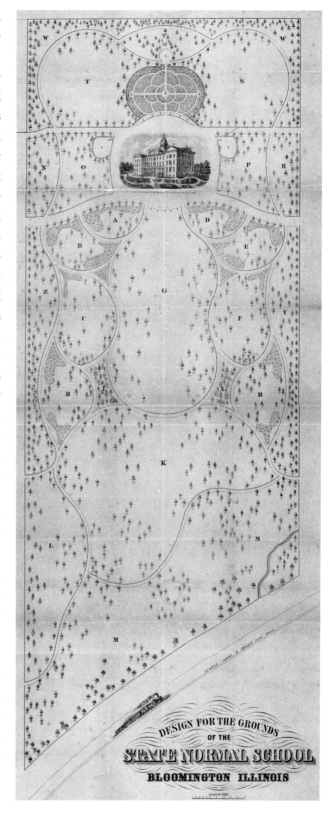

ture Nelson was a little-experienced pioneer, not a professional; Bryan's patronage had perhaps launched his American career. The scale and complex scope of the cemetery project may have exceeded his abilities, making more expert assistance necessary, and Graceland's 1861 charter confirms Nelson's role as a subsidiary one, listing him as "Assistant Landscape Gardener" to Saunders.[26] The desire to attract prestige to Graceland may have been another motivation behind Bryan's move to secure a nationally eminent designer.

Philadelphia was a logical place for Bryan to seek assistance. That city had already become, as Aaron Wunsch has shown, the "capital of cemetery production."[27] Not only was Philadelphia "replete with garden-like graveyards" such as Laurel Hill, "it also hosted enterprises like Robert Wood's iron foundry, John and Matthew Baird's marble works, and Robert Buist's nursery—sources of goods that graced cemetery lots from New York to New Orleans, and beyond."[28] And of course it was the host city of John J. Smith's *Horticulturist;* Smith often promoted Saunders in the magazine, and Saunders was a regular contributor.[29] More immediately, Bryan may have taken his cue from Rosehill Cemetery; by March 1859, as he already knew, Rosehill had retained Saunders at Smith's suggestion.

After having begun some work for Bryan (he received a check in January), Saunders visited in April 1860 to meet Bryan and survey the cemetery site in Lake View.[30] On the eve of his return to Philadelphia, Saunders recorded his impressions of the property in a letter dated April 4, 1860; the only letter from the landscape gardener to his client known to exist, it is worth quoting in full:

Dear Sir,

In complying with your request to express, in writing, my opinion with reference to the most desirable disposition of the grounds and *properties belonging to Mr. Healy,* I unhesitatingly state that I consider it unequalled for the purposes of a cemetery, by any grounds within the same distance of Chicago, so far as I have seen. It possesses every requisite so far as accessibility, configuration of surface, elevation, and existing vegetation are of importance. This applies more particularly to that portion of the property lying eastward of the public road. Although not so elevated, yet the grounds west of the road are for the most part available for cemetery purposes, if found necessary to employ them for that use. The soil in these grounds is rather deficient in the qualities essential to the successful growth of ornamental trees, but, in connection with your own grounds which are contiguous, and abound in the richest soil, a mutually advantageous improvement might be effected. *Your grounds* prevail in that expression which is usually termed park-like, suggestive of rural, domestic homes, qualities which, of course, rather increase than diminish its adaptability to cemetery purposes, which would undoubtedly be the best disposal of the property, provided that adjoining was so appropriated. To establish such a cemetery, an expenditure of from 15 to 20,000 dollars would be required to open roads, locate lots etc so as to bring it into a saleable condition 5 to 6,000 dollars would suffice other and further improvements being left for the future. I am very well convinced that the above properties would form a cemetery that could not be excelled for natural beauty by any grounds within the same range of the city.

Yours very respectfully,

William Saunders[31]

This is a document rich in revelation. At the outset, we learn that Saunders is assessing the suitability of not one but two properties for cemetery use, Bryan's and a second owned by his Bird's Nest neighbor, the painter George Healy. As the impetus for the report was Bryan's request for an evaluation of the artist's grounds, not his own, Saunders had presumably surveyed Bryan's on his earlier visit to Chicago in March. Having now inspected them both firsthand, he proposes that the contiguous properties be combined to form the cemetery. As Saunders made his fulsome endorsement against the backdrop of his ongoing work for Rosehill, either the collective tract was indeed remarkable or he was especially keen to secure this new job. This letter also points to Healy as the first to join Bryan in the venture, and in fact, earlier that year, in January, Bryan had persuaded the artist to purchase the eighty acres adjoining his own.[32] As Saunders identified the properties by their owners' names, Bryan apparently had yet to form a corporate entity. Perhaps Saunders's cost estimates now became one reason to do so.

For Bryan, however, the cemetery was not progressing quickly enough. With the suitability of the properties now confirmed, he accelerated the project. On April 6, 1860, only two days after Saunders's report, Bryan publicly revealed and promoted the cemetery, now named Graceland, in a newspaper advertisement. "This new rural burial ground," Bryan announced, "has just been critically inspected by William Saunders, Esq., the eminent Landscape Gardener, who was called from Philadelphia to decide for the owners, whether to devote the land to villa sites" or to a cemetery.[33] This advertisement is the only document to suggest that Bryan had considered an alternate use for the property, although this may have been a marketing fabrication to imply that the land was beautiful enough to attract homebuyers. Bryan then continues with an extensive quotation from Saunders's endorsement of the tract's suitability as a cemetery site. In the advertisement Graceland Cemetery Company is identified as the owner of the land. And yet in Saunders's report, dated only two days earlier, Bryan and Healy are named as the owners. Perhaps Bryan had been poised to act the moment he received favorable word from his landscape gardener. Presumably having prepared the necessary legal instruments in advance, Bryan apparently purchased Healy's parcel, established a corporate identity, and then transferred the land title to it within only forty-eight hours. Although Bryan identified himself as the cemetery company's president, at this point he may well have been its only stockholder.

With the Graceland Cemetery Company now in existence, Bryan became fully immersed in the project, "greatly neglecting [his] other business." Writing in May to placate a Virginia investor frustrated by Graceland's delayed start, he explained: "I rise by dawn, and travel twenty-four miles to the Cemetery every morning (Sundays excepted), before other people are generally out of bed. I remain on the ground often all day without stopping for dinner, and for any personal services." This letter also confirms that Graceland was proving costly. Up until then, the enterprise had required him to hire, among others, "a professional landscape gardener at a salary," although he had "already paid $250 for gardener's services," and Saunders, Bryan lamented, had "scarcely begun." He also remarked that the cemetery's "roads [had] cost thousands of doll[ar]s!" As Saunders had not yet completed his design, the roads Bryan mentions must have been the product of Swain Nelson's work. Bryan also anticipated further expenses, such as the design and construction of a chapel, an office, a "Superintendent's dwelling," and a "stone archway at the entrance."[34]

THE GRACELAND CEMETERY COMPANY

By June 27, 1860, Bryan's efforts to secure partners in his cemetery venture had met with success, for on that date he publicized the Graceland Cemetery Company's "Organization & Election of Officers."[35] If Bryan's April advertisement was accurate, the company was already some two months old, so "organization" here may refer to its legal incorporation. "On Monday last," Bryan announced, he had been elected president and an unnamed individual (Healy, according the cemetery's charter) as treasurer.[36] William B. Ogden, Edwin Holmes Sheldon, and Dr. Sidney Sawyer were elected the company's board of directors (later known as the board of managers). The new company's office was in Bryan Hall.

Apart from its members' mutual financial interest in this business enterprise, Graceland's officers and its board were already a long-standing circle of friends sharing other common bonds. One of the foremost was a preoccupation with Chicago's civic improvement. William Butler Ogden (1805–1877) had come from New York to the frontier outpost in 1835. Amassing a fortune from land speculation and infrastructure projects such as canals and railroads, he had also served as Chicago's first mayor (1837–1838). As one of his biographers put it, Ogden believed that "anything that benefited Chicago benefited him."[37] In 1837, for instance, he lured the architect John Mills Van Osdel (1811–1891) from New York with the promise of abundant commissions, including one from him. When Van Osdel took up the offer the city gained its first architect, and its former mayor soon erected Chicago's first architect-designed residence. Ogden's similar promise to Healy led him to abandon his Paris studio for Chicago in 1855.[38] Although other painters were already living in Chicago, the art historian William H. Gerdts notes that "an artistic community did not begin to form" until Healy's arrival. "Already renowned as a portraitist of international acclaim," Gerdts writes, Healy "provided Chicago with sophisticated cultural identification."[39]

Less is known about founding board member Dr. Sidney Sawyer (1810–1894). His brief obituary in the *Chicago Tribune* noted that Sawyer, a medical doctor, had come to Chicago in 1838 and "established a drug business, which was conducted so successfully that he retired a wealthy man."[40] We also know that he was married to one of the daughters of Justin Butterfield, from whom Bryan purchased the land for Graceland Cemetery.[41] A historical account of the city's pharmacists published in 1903 notes that Sawyer's store specialized in the selling of patent medicines (notably Sawyer's Extract of Bark, recommended for fever and ague) and that Sawyer was "elected health officer" in 1849, which may well have been the source of his interest in the new cemetery.[42]

Along with civic concern, an interest in landscape gardening also apparently united the board's members. As we saw in chapter 1, following Bryan's building of Bird's Nest, George Healy had also made his own picturesque Cottage Hill retreat, Clover Lawn. William Ogden, too, had earlier cultivated a remarkable estate within the city itself. When Healy first came to Chicago,

2.5. William B. Ogden's house and garden around 1850. The building perished in the Great Fire.
Courtesy Historic Architecture & Landscape Image Collection, Ryerson & Burnham Libraries, Art Institute of Chicago.

he lodged with Ogden, who was a bachelor. Ogden's Van Osdel–designed house was actually a double one; Ogden's sister and her husband, Edwin H. Sheldon (1821–1890)—also Ogden's business partner and later fellow Graceland cofounder—resided in the other half. The dwelling occupied an entire bucolic, garden-filled block. (Fig. 2.5) Decades later, when Healy wrote his memoirs, Ogden's place lingered still in his memory. The artist recalled it as "a large, roomy, comfortable, old-fashioned frame house, spreading broadly in the midst of the enormous garden, or 'yard,' as people modestly called their gardens, —the trees were superb, the flower-beds of the brightest hues, the lawn stretched before the house: it was a delightful residence, a town house with the pleasant aspect of a country place." For Healy, Ogden's home "well justified" Chicago's "Garden City" moniker.[43]

Landscape gardening would have held special interest for an artist such as Healy. In the cemetery founders' day, the pursuit was often seen as analogous to making three-dimensional outdoor landscape paintings. Although renowned as a portraitist, Healy also painted landscapes, and his lesser-known works in this genre are known to have included gardens.[44] Yet even Healy's portraits can be considered analogous to landscape designs. Rather than valuing every detail equally and attempting to replicate them exactly, Healy conceived portraits as a whole and paid close attention to the visual effects of, for instance, shadows and color contrasts. Not unlike a landscape garden, his portraits are in their own way just as contrived, and yet they appear naturalistic.[45]

This board of landscape gardening aficionados was probably already acquainted with the new rural cemetery type. Healy, a native of Boston, had opened a studio there at the age of eighteen, in the autumn of 1831, the same year that saw the well-publicized opening of America's first rural cemetery, Mount Auburn, in nearby Cambridge.[46] Bryan may also have become familiar with the cemetery when studying at Harvard in the late 1840s; he would promote Graceland as the "Mount Auburn of Chicago."[47] He also undoubtedly knew Cincinnati's Spring Grove Cemetery in the early 1850s, when he was practicing law in that city; and Healy may have known Père Lachaise Cemetery in Paris. William Ogden, too, was no stranger to this new type of burying ground: he had cofounded Rosehill Cemetery the year before.

THREE

The Earliest Designs

No firsthand textual descriptions of William Saunders's Graceland design have survived, but two reports he wrote around the time he obtained the Graceland commission suggest his design approach. The first, an account of his plan of Hunting Park, appeared in the October 1858 *Horticulturist*, only a few months after his Chicago stay with Bryan at Bird's Nest. In his layout for this park in Philadelphia, Saunders noted, he had "not attempted to produce intricacy by an arrangement of tortuous or abrupt curving walks, but the various groups will be planted sufficiently thick, and intermixed with appropriate undergrowing plants, so as to produce a fresh change of scenery at every step, and thus avoid tameness of expression."[1] This description suggests the fluidity of Graceland's layout, and it offers an insight into the nature of the planting and the spatial qualities he may have envisioned for the cemetery. The second report, dated February 12, 1859, accompanied his unsuccessful entry in the competition to design Fairmount Park, also in Philadelphia. Written the same year Saunders secured the Graceland job, the document sets out his design objectives for the new park:

1. The preservation of the natural beauties of the ground, existing trees and other vegetation.
2. To provide a sufficient number of roads and walks, both for carriages and pedestrians, from which the beauties of the grounds and surrounding scenery may be observed.
3. To arrange additional plantations so as not to obstruct de-

sirable views; to form combinations varied and pleasing, both in color and outline; and at the same time provide all desirable shade and shelter without detracting from these important specialities.

4. To form a design capable of further embellishment in the future, without detracting from its present or future merits.

5. Economy of construction.[2]

It is likely that Saunders had the same general aims in his design for Graceland, but for most of the specifics of his plan for the cemetery we must turn to the only image of it known to have escaped Chicago's Great Fire of 1871. This undated lithograph, titled "Graceland Cemetery, being Subdivision of the South Part, East of Green Bay Road, SW ¼ Section 17 T40N R14 E," is the earliest known representation of the cemetery's initial layout.[3] (Fig. 3.1) The plan records "Wm. Saunders of Philadelphia" and "Swain Nelson of Chicago" as the designers and "E[dmund]. Bixby" and "S[amuel]. S[ewell]. Greeley" (1824–1916) as the site's surveyors. It was made by the well-known Chicago lithographer and engraver Charles Shober, whom Graceland probably commissioned to produce the document as its foundation or record plan.[4] Although the plan attributes the cemetery's layout to both Saunders and Nelson, there is no evidence that the two actually collaborated. Most likely Saunders, after inspecting the site and presumably meeting with Nelson, returned to Philadelphia and produced the plan there, and Nelson then implemented it.

In his layout, Saunders organized the cemetery with a fluid, curvilinear network of carriage roads and paths, beginning at the main entrance at the junction of Green Bay Road (now Clark Street) and Albert Street (now Irving Park Road). As we saw in chapter 2, the decision to configure the main entry at the southwest corner of the property was apparently made before either Saunders or Nelson became involved. Saunders did, however, provide a secondary entrance further north, off Green Bay Road, to enable access to the stables that were to be built across the road—prompting us to remember that teams of horses, not motorized vehicles, were used in maintenance operations. In keeping with rural cemetery design tenets, Saunders's sinuous thoroughfares contrasted sharply with the city's ubiquitous, unrelenting street gridiron. For Chicagoans, one imagines, a carriage ride through Graceland would have

(*Opposite*) **3.1. The foundation plan of Graceland Cemetery by William Saunders and Swain Nelson (c. 1860).** Courtesy Chicago History Museum (ICHi-51565).

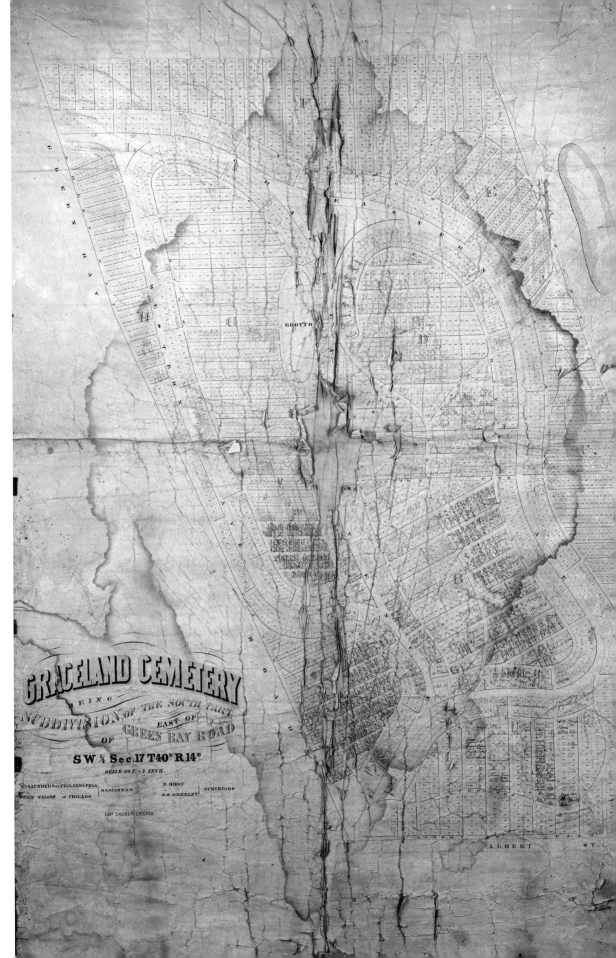

GRACELAND CEMETERY

BEING

SUBDIVISION OF THE SOUTH PART EAST OF GREEN BAY ROAD

SW ¼ Sec. 17 T 40 N R 14 E

SCALE 60 F! = 1 INCH.

W.M SAUNDERS OF PHILADELPHIA — DESIGNERS — E. BIXBY
SWAIN NELSON OF CHICAGO — S.S. GREELEY — SURVEYORS

CHAS SHOBER, CHICAGO.

GROTTO

ALBERT ST.

been an ethereal experience. Within this frame of roads and paths Saunders allocated fifty acres for burials, dividing them into twelve sections, alphabetically labeled from A to L. These were then subdivided into thousands of individual grave plots.

Beyond its function as a private cemetery, Graceland would also serve as a quasi-public park; indeed, rural cemeteries are today considered precursors to public parks. Presumably following Bryan's instruction, Saunders reserved thirty-six acres at the northern end of the property as parkland. At Graceland, then, the park and the rural cemetery converge. It is unclear, however, whether Bryan intended this portion of the grounds to remain parkland, as it would later be used for burials.

Saunders's layout included another expansive parklike feature intended for the living: a grotto within section G, near the center of the cemetery, an elliptical area that was not divided into burial plots. Grottos go back at least as far as classical antiquity; for the ancient Romans, these subterranean vaults were shrines to the gods and the mysterious dwelling places of water nymphs. But they also have a long history as garden features, appearing first in the gar-

3.2. The grotto (1748) at Stourhead landscape garden, Wiltshire, England. Photograph by John Tatter.

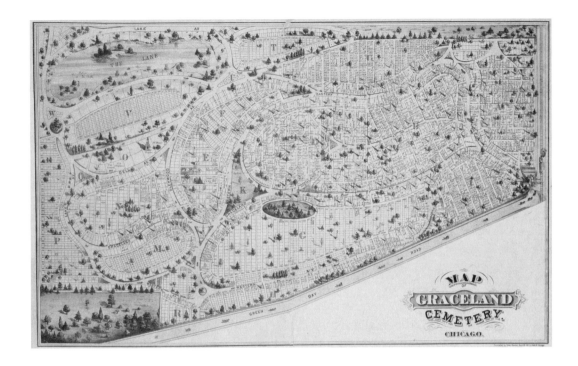

3.3. Charles
Rascher's map
of Graceland
Cemetery (c. 1878).
Courtesy Chicago History
Museum (ICHi-27725).

dens of Renaissance Italy and later in France and England.[5] (Fig.
3.2) Often emulating natural caves, they were usually constructed
of rough-hewn stone and enhanced with cascades, fountains, or
other water features, and they served as cool retreats from summer
heat. They often included sculptures of mythological figures, and
they were placid spots for contemplation and reflection. By the mid-
nineteenth century grottos had passed from fashion, so the decision
to include one at Graceland is somewhat surprising.[6] It is perhaps
evidence that Saunders and Bryan shared an appreciation for the
English landscape sensibility, and possibly they found the symbolic
association of the grotto with life and death compatible with a cemetery setting.

A map by the Chicago cartographer Charles Rascher, dating
to around 1878 (see chapter 4), is the only other plan to document
the grotto. (Fig. 3.3) In this image, rather than the discrete cave-
like construction we might anticipate, the grotto is rendered as a
picturesque parkland, replete with lawns and groves of evergreen
trees.[7] But Rascher's map, like Shober's lithograph, does not represent landforms and gives no hint of any fountains, rockwork, or
subterranean features. Beyond this, we know only that at some
point before 1886 the grotto parkland was sacrificed to create more
burial space. Perhaps the cost of maintenance and security grew

too high, or perhaps the well-drained site could not retain water. In a Graceland plan dated that latter year, the area is shown as subdivided into plots and renamed the Vault Section, to accommodate family mausoleums.

In the lower-lying northeastern area, Shober's lithograph depicts a water body of some kind, indicated by a nearby walk called "Lake Path." Given its somewhat amorphous outline, possibly it was a naturally occurring slough or swamp. The area around it is not shown as being subdivided into burial plots; it is simply blank. It is possible that the lake or slough then lay beyond the cemetery's boundary, or perhaps the map merely indicates that the area is not yet available for burials. Saunders had earlier designed an artificial lake at Rosehill, and he might well have envisioned a similar transformation here, if only as one of the "further improvements being left for the future."[8] Graceland was already proving to be an expensive venture, and such a costly proposal may have been postponed, since there was no urgent need for such an ornamental feature. Still, in 1862 the *Chicago Tribune* listed "miniature lakes" as among the improvements Saunders had orchestrated at Graceland, so what this section of Shober's lithograph actually indicates remains unclear.[9]

The lithograph does not show existing vegetation or proposed new plantings, and neither Saunders's own drawings nor any "accompanying references and remarks explanatory" have come to light. This is an especially disappointing loss, as Saunders claimed that the "careful indication" of the "position of every tree and shrub" distinguished his plans from those of his competitors.[10] Perhaps his design was largely confined to laying out drives and paths and the requisite earthworks. A contemporary newspaper report notes that the site is covered with groves of "patriarchal" oaks, which might have diminished the need for extensive planting (apart from turf), at least in the short term, and lent the new cemetery a pleasingly antique patina.[11] Bryan's workers were confronted with the task of "subduing a wilderness of underbrush";[12] nonetheless, documentary evidence suggests that new plantings were among the initial improvements. In May 1860, for instance, Bryan recorded purchasing evergreens, and he foresaw the need for "other trees and plants."[13] The names Saunders gave to some of Graceland's walks also offer clues to kind of vegetation that was at least anticipated, if not yet in place. As recorded in Shober's lithograph, these include "Shady," "Shrubbery," "Leafy," and "Evergreen" paths.

THE FIRST BUILDINGS

Swain Nelson recounted that Bryan first secured plans for the cemetery's "entrance and office" in 1858 or 1859, before involving any landscape gardener. Bryan described the plans in a letter to one of his Virginia real estate clients in May 1860, reporting that "a chapel has to be built, to answer also for office," along with a "stone archway at [the] entrance."[14] The next month, the "arched entrance gate and adjoining buildings of stone" were under construction."[15] As this structure was razed around 1896, our appreciation of it must rely largely on textual sources and a few surviving images. In August 1860 the *Chicago Tribune* described the two-story edifice, evidently complete by then, as a "massive and characteristic [i.e., local] stone structure" that included "large waiting rooms on either hand."[16] Graceland's "broad entrance," a later report said, was an "imposing" gateway, "arched between buildings on either side," with one side accommodating "a chapel and office" and the other a "vault or public receiving tomb."[17] (Fig. 3.4) Rather than erect a collection of individual structures, Bryan had chosen to combine chapel, office, and vault into a pair of blocks.

3.4. The cemetery's original entry gateway around 1893. Completed by August 1860, the structure accommodated a chapel, office, and receiving vault; it was razed around 1896.
From Simon, *Chicago, the Garden City.*

These were positioned opposite one another on each side of the entry drive and linked overhead by an arch to form the gateway. The *Tribune* later described the chapel and waiting room as "free from that dreary look which at such places seems premonitory"; instead, Graceland's building was "positively cheerful," replete with "spiritual-looking pictures on its walls and a canary singing blithely in its cage."[18] Soon "a capacious and an elegant greenhouse" would be added to the side of the chapel, for lot holders who wished to winter over plants.[19] The few known images of this architectural ensemble reveal it to be a Gothic Revival structure, a style then popular owing to its compatibility with picturesque aesthetic ideals. Its ecclesiastical connotations also made it especially appropriate for cemeteries. When viewed against the sylvan backdrop of the broader cemetery, the gateway structure was as much a landscape garden feature as it was an isolated building.

Beyond this impressive entrance, in what amounted to the cemetery's first preliminary layout, according to Nelson, the gateway plan also projected "a straight road [configured on a diagonal] to a given point." It is this layout that first fixed Graceland's entrance at its southwest corner, near the intersection of the former Green Bay Road and Albert Street. Incorporating the diagonal segment of the entry drive at the head of his Main Avenue, Saunders's new layout retained this original entrance.

It is not known who designed the entry buildings and archway, but it is likely that Bryan commissioned a local partnership, architects Asher Carter (1805–1877) and Augustus Bauer (1827–1894).[20] By 1859 Bryan had erected at least four buildings in Chicago, ranging from frame dwellings to commercial buildings such as Bryan Hall, then under construction, and for each of these he engaged Carter & Bauer as his architects.[21] In 1866 he commissioned Bauer, now practicing alone, to design a new vault for the cemetery.[22] And around 1869, no doubt on Bryan's recommendation, his brother-in-law Jedediah Lathrop hired Bauer to design additions to Huntington, his mansion at Elmhurst.[23]

After John Van Osdel, Asher Carter was Chicago's senior architect.[24] A carpenter's son, Carter, like Van Osdel, was self-taught. He came from New Jersey to Chicago in 1849, initially to superintend construction of the Second Presbyterian Church. This church was, for at least one professional journal, "the first building of any architectural pretensions erected in Chicago."[25] Subsequently he established a local practice, and he formed a partnership with Augustus

Bauer in 1855. Bauer was professionally trained, a graduate of the Kunst- und Gewerbeschule in Darmstadt in his native Germany.[26] In 1859 Bryan commissioned the pair to design Bryan Hall, and he instructed his "excellent architects" to "make the building plain and unostentatious, but solid and permanent."[27] Indeed, these qualities accurately characterize Carter & Bauer's entire known oeuvre and were likely what attracted Bryan to the firm in the first place. In a short tribute written a the time of his death, a colleague noted that Asher Carter did not have a "talent for elegant embellishment" because "his tastes were too severe, his sentiments too fastidious, to indulge in imagery."[28] Similarly remembering Augustus Bauer's works as "largely of a utilitarian nature," his colleagues in the Illinois chapter of the American Institute of Architects wrote that he seldom "distinguish[ed] himself by brilliancy or originality of design." Yet although Bauer "never indulged in any extravagance of design," they continued, "it can truly be said that nothing that is commonplace was ever produced by him."[29]

Bauer's German professional qualifications, complemented by his New York credentials, undoubtedly appealed to Bryan's sophisticated taste. He also possessed, according to his peers, a certain "element of honesty and straightforwardness which attracted clients." Never deceptive about the probable cost of projects, "he estimated liberally, knowing that he would always require good work that costs more than bad work, and no one was ever disappointed in the result."[30]

Along with its significance as a component of Graceland's initial design, the entry building also helps us establish the cemetery's early chronology. If, as convention holds, the cemetery dates to some unknown point in 1860, then this "massive" and "imposing" structure—completed by the end of August that year—was commissioned, designed, and built within eight months, or even less if we assume that little construction activity took place in the winter months. This unlikely scenario is another factor that points to 1858 or 1859 as the year of Graceland's origin and an indication that Swain Nelson's memory was probably accurate.

IMPLEMENTING THE NELSON–SAUNDERS DESIGN

In its account of Graceland's dedication, the *Chicago Tribune*, presumably with Bryan as its source, reported that the cemetery's "improvement commenced in April [1860]."[31] That month also saw the first burial, when, on the thirteenth, the remains of Bryan's infant son Daniel were reinterred from City Cemetery.[32] If the implementation of Saunders and Nelson's layout began in April, its progress was apparently swift. Only a month later the newspaper reported that "nothing could be more attractive than [Graceland's] internal arrangements, or its surroundings." Although it did not name a designer, the *Tribune* attributed the cemetery's layout to "a gentleman well skilled in the art."[33] Still, in May 1860 many of the improvements were undoubtedly yet to be executed. Graceland was dedicated on August 30, 1860, and the *Tribune*'s report on the event offers further insights into the status of the implementation of the design. The newspaper now identified both of Graceland's designers, "Wm. Saunders, Esq., the eminent Landscape Gardener, of Philadelphia, [and] Swain Nelson, of this city." It noted that "the grounds are enclosed [with wooden fences] and the broad gravelled roads and winding paths among the trees are all finished." The stone gateway entry was also complete. There is

(*Right and opposite*)
3.5. and 3.6. Burial plots at Graceland around 1870; note the stone curbing and sand footpaths. The cemetery would later prohibit curbing, and some paths were eventually covered with macadam.
Courtesy Chicago History Museum (ICHi-61232 and ICHi-23338).

no mention, however, of new plantings. The same article stated that "quite a number" of burial plots had "already been purchased and improved." Presumably "improved" refers to laying sod and the traditional practice of bounding a lot with stone coping. (Figs. 3.5 and 3.6) Illustrating the influence of Laurel Hill, Graceland allowed owners to enclose their plots only with hedges, "nothing in wood or iron." Such natural enclosures would, the *Tribune* said, "secure a peculiarly charming effect to the improvements of the grounds."[34] Notions as to what constituted "charming effects" would change, however, and eventually hedges and even stone coping would be prohibited. Despite the newspaper's attention to the improvements already in place, it is highly unlikely that Nelson had fully implemented the design between April and August 1860. Realizing a major landscape initiative such as this would have consumed years, not months, and undoubtedly the report exaggerates the status of the work.

In February 1861, some six months after the dedication, the General Assembly of the State of Illinois passed a bill titled "An Act to Incorporate the Graceland Cemetery Company."[35] The perpetual charter it granted was a very liberal one; the cemetery company was given police powers, for example, and it was granted the right to expand to up to five hundred acres. It was also empowered "to sell, exchange or dispose of any part or parcel of land that [it]

may be compelled to purchase in order to obtain such grounds, as may not be actually need[ed] for burial purposes." The charter also exempted Graceland's property from potential state or municipal condemnation, and, of no less import, it was not subject to real estate taxes. At the time of its incorporation, portions of the company's property were still "in the process of tasteful improvement."[36] The company first took advantage of the clause allowing it to expand in 1861, when it purchased an additional "45 acres to the west" of Green Bay Road from Conrad Sulzer.[37] A portion of this newly acquired land would become the site of the cemetery's stables, and the rest was used for residential developments.

The year 1861, when the Civil War began, provides us with a glimpse of the city through the troubles of one of Graceland's managers, the painter George Healy. By June, less than a year after the cemetery's dedication, Healy was in considerable personal financial difficulty, exacerbated by the outbreak of the war.[38] As he had discovered when he arrived in the city in 1855, "land agents, merchants, and bankers were more plentiful than artists in Chicago."[39] In his memoirs he wrote that he soon succumbed to "the 'land fever' then raging" and gained a "well-earned reputation for bad speculations."[40] Healy's daughter later vividly recounted her father's plight, one no doubt shared by others, and the speculative climate that pervaded Chicago in the 1850s:

This was a period of prosperity and wild activity. Stories were afloat of men who had bought a tract of land and sold it again, almost immediately, at enormous profit. Of those who lost everything instead of making a fortune, less was said. Even an artist, so ill-fitted by nature and education to understand anything about business, cannot always escape the fever of speculation. The painter from over the seas caught it pretty badly. In that first year [1855–56] he had been so very successful that probably he felt justified in risking some of his hard-earned dollars—and he bought right and left, trusting to any and every adviser, accepting land, which he never seems to have even seen—had he time for such details?—in payment for his portraits. Some of these purchases, advised by his real friends, proved to be good investments. Others, thrust upon him by pretended well-wishers, were disastrous. One lot was discovered to be covered by the lake. And, so, much of his gains of that first hard year were literally washed away.[41]

Healy's real estate misfortunes compelled him to withdraw from the Graceland Cemetery Company that October. It was Healy's patron who came to his aid, as the appreciative artist reported to a friend: "Our Lord has helped me by inducing my neighbour Thomas B. Bryan to offer to take upon himself $18,500 of my $23,000 debts for my interest in Graceland Cemetery. I have no doubt this property will be in five years worth one hundred thousand dollars, and yet I rejoice in his good fortune, and my relief."[42] Edwin H. Sheldon, the business partner and brother-in-law of William B. Ogden, now replaced Healy as the company's treasurer.

In December 1862 the *Chicago Tribune* again returned to Graceland and surveyed the "improvements completed and in progress," touring the grounds with the cemetery's superintendent, Elias Olson.[43] The report opens with the boast that the newly expanded cemetery is larger than its Philadelphia counterpart, Laurel Hill. Saunders's "skill and taste" are praised: "He has displayed a happy faculty in availing himself of the varieties of the surface, for the introduction of serpentine roads, miniature lakes, and the like." The writer's enthusiasm apparently undiminished by the frigid winter weather, he continues: "The extent of thorough and permanent improvement, considering the fact that the cemetery is quite in its infancy—not yet two years old—is really wonderful, and is an earnest of the high rank Graceland must ere long assume among American cemeteries." Many of the individual burial lots, he reported, were already adorned with the "fine stone curbing" that the cemetery later prohibited on aesthetic grounds. (Figs. 3.7 and 3.8) And, despite the ongoing Civil War, the article notes that there are "many other fine improvements contemplated and contracted for, which will be completed during next season." Presumably the superintendent, Elias Olson, was the source of this information, although it may have been Bryan himself. The story concluded with a prophecy: "Graceland promises to become as famous for the architectural skill and elegance of its artificial adornments [that is, grave monuments], as it already is for its natural beauty."[44] More a promotion than a critical assessment, this report must have pleased Bryan with what amounted to a free advertisement. Accounts such as these, however, must not be dismissed merely as crass gratu-

3.7. This photograph, taken around 1870, shows Graceland's original wooden perimeter fence.
Courtesy Chicago History Museum (ICHi-61233).

ities; they register the burgeoning city's enthusiastic embrace of the future and its cultural possibilities.

Graceland undertook its second expansion in 1864. Shifting direction from its earlier expansion westward, it now acquired "35 acres to the east" of Green Bay Road.[45] By then, the Chicago *City Directory* could distinguish the cemetery's lots as "well-sodded," each "surrounded by a coping of Athens marble, eighteen inches high" and "a cedar hedge," features that lent "a unique and tasteful air to the grounds." And, in one of the earliest references to Graceland's planting, the entry also noted a "profusion of ornamental shrubbery" at the cemetery.[46] In July of that year, the *Tribune* again overviewed Graceland's continuing development, this time at greater length. Rather unusually, the article begins with a substantial meditation on death, burial, and the afterlife, a tour-de-force that touches on the ancient Egyptians, Pythagoras, the Gospels, the Koran, and Shakespeare, among others. In addition to the requisite descriptions of the landscape and major monuments, the second part of the report also describes the emotional and metaphysical qualities of the setting:

> There is a peculiar charm about Graceland; there is a settled calm, a quiet solitude in the air which awes the soul,

saying "Peace, be still!" It has so much of the picturesque—both natural and artificial—as to cause the evolution of devotional meditation and make that agreeable which in itself is usually regarded as repulsive. It is well shaded by trees over the whole extent, and its meandering walks seem to invite the visitor to lose himself in a labyrinthine maze of thought as of bodily location. The grounds are divided into lots, not squarely as is the case with many cemeteries, nor fancifully to the abuse of diversity; there is just enough of the absence of uniformity to avoid fatigue on the part of the visitor, though he should ramble for hours within the enclosure.

3.8. Another photograph from the same period shows the mausoleums (*right*) and the carriage drive. Courtesy Chicago History Museum (ICHi-61234).

The writer notes that much of the success of the cemetery's landscape effects can be attributed to the prohibition of walls and fences to enclose lots—"a wise provision, morally as well as for the sake of ocular symmetry," as "the graveyard is the last place for the exhibition of that exclusiveness which sometimes obtains in our little social world." For the Christian believer it should matter little where his bones are laid, "yet there is something attractive in the scene, which amid all the repulsiveness of death whispers, 'Here is a holy calm retreat.'"[47]

FOUR

A Decade of Expansion

In April 1865, back east in Thomas Bryan's native Virginia, Robert E. Lee surrendered on behalf of the Confederate forces after four long years of war. Only six days after peace descended, however, Abraham Lincoln's assassination violently replaced jubilation with grief. The nation had lost its leader, Illinois its adopted son, and Bryan a personal friend. So close was their friendship that he was awarded the privilege of serving as a pallbearer both in the Chicago funerary procession on May 1 and at the martyred president's final funeral service in Springfield on May 4. Lincoln was interred at the Oak Ridge Cemetery in Springfield, designed in the late 1850s by William Saunders, one of Graceland's first landscape gardeners. In the coming months, Saunders would lay out the tomb's six-acre surrounds and position the National Lincoln Monument within them.

Although his home state had joined the Confederacy, Bryan was "ardently pro-Union," holding the view that "Southern leaders were 'arch traitors, alone responsible for the war.'"[1] During the war, Bryan patriotically assumed a leadership role, "organizing forces, and providing for them in the field"; he presided over the Northwestern Sanitary Fair of 1865, which was held in Chicago and raised over $300,000 in war relief funds.[2] He also spearheaded the effort to construct a soldiers' home in the city and served as its president.[3] In that capacity, "he purchased and gave to his wounded veterans the original copy of the Emancipation Proclamation," which enabled them to raise funds by selling lithographed copies.[4] Bryan's initiatives did not escape President Lincoln's attention. Moreover, in recognition of his "faithful services in maintaining the honor, integrity and supremacy of the government of the United States," Bryan was later elected, by unanimous vote, to the Military Order of the Loyal Legion of the United States.[5]

More broadly, the Civil War was, as the historian Theodore J. Karamanski observes, "a crucial event in the development of nineteenth-century Chicago," and he cites the Union Stock Yard, which opened on December 25, 1865, as a symbol of the effect of the war on the city's growth:

> The war directed the flow of vital food commodities away from Chicago's most persistent urban rivals, which were too close to the front lines during the first two years of the war and were hurt by stoppages of trade on the Mississippi Rivers. Because the war cost St. Louis its status as the major grain distribution center and Cincinnati lost its distinction as the pork-packing capital, Chicago emerged as the logical center for the meatpacking, wheat distribution, and related industries. Heavy industry took root in Chicago during the war to provide Union forces with the rolling stock and rails needed to transport troops and supplies."[6]

Chicago was also home to one of the Union's largest prisoner-of-war camps, Camp Douglas, which at times held more than ten thousand Confederate soldiers. The mortality rate among the prisoners was extremely high, in large part because of poor sanitation and lack of medical care. By the war's end, the city itself had lost nearly four thousand men.[7] Some of these casualties, including at least one who "died in rebel service," were interred at Graceland.[8]

THE GRACELAND CEMETERY IMPROVEMENT FUND

In December 1862 the *Chicago Tribune* reported on what it considered the "fatal defects" of Rosehill and Graceland cemeteries, its scrutiny ostensibly attracted to the pair by their beauty and Chicagoans' nascent pride in them.[9] The problem lay in the cemeteries' charters. As long as Graceland and Rosehill "have lots to sell," the newspaper argued, "it is for their interest to keep their improvements in good order." But what would happen, it asked, after all the lots were sold? Who would maintain the cemetery then? Could "Mr. Bryan" be expected to "incur an annual charge of three thousand to five thousand dollars for doing that which will yield no return?" Neither charter included a provision "compelling the proprietors, as they sell lots, to place a certain percentage of the proceeds of each sale in the hands of trustees for investment," with the income specifically earmarked for maintenance of the grounds.

A few days later, however, Bryan revealed that when he "conveyed the land to the Cemetery Company," the deed included a provision for a fund devoted to maintenance—"ten percent of the gross receipts"—and he wrote that although the board of managers had entrusted it to his management, he would "most gladly transfer to other hands the expenditure of the fund, say to a committee appointed by the lot owners, and acting entirely independently of the managers and myself."[10] Thus it was that concern for the "perpetual care of the grounds" led the Graceland Cemetery Company to revise its charter, and in February 1865 the General Assembly of the State of Illinois passed an act to incorporate the Trustees of the Graceland Cemetery Improvement Fund.[11] This act vested "ten per cent of the gross proceeds of all sales of lots" in the trustees (who were to be elected by the lot owners themselves), to be expended on "the improvement, ornamentation, preservation and maintenance of the grounds, walks, shrubberies, enclosures, structures, monuments and memorials, and any and all other things in and about" Graceland.[12]

SWAIN NELSON AND LINCOLN PARK

As we have seen, the city's decision to cease lot sales in City Cemetery and relocate the existing graves had prompted and quickly galvanized the formation of Chicago's first rural cemetery, Rosehill, in 1859. And, although Thomas Bryan had already been planning to develop a cemetery for some years, the actions of the city's Common Council were also an impetus for the founding of Graceland. By 1865 Rosehill and Graceland had both received disinterments from City Cemetery, and more were about to arrive. That year, the Common Council ordered that the old cemetery be vacated, "authorizing lot-owners to exchange their lots for lots in any of the new cemeteries, of equal size and of their own selection."[13]

In 1865 the city was poised to convert an unoccupied northern section of City Cemetery into parkland. Positioned alongside Lake

Michigan just north of the city, the fifty-acre tract had been designated Cemetery Park two years earlier. But apart from changing
its name to Lake Park in 1864, the city took no immediate action
to actually transform the sandy expanse, and for nearly three years
the park existed only on paper.[14] The *Chicago Tribune* lamented,
"We have parks in abundance on our maps, the names of which
sound well, but when we examine their area, which is seldom more
than a single block, our ideas of the parks of Chicago 'grow small
by degrees and beautifully less.'"[15] In 1865 the city renamed the
park yet again, this time in honor of the martyred president, and
at last appropriated $10,000 for its improvement.[16] Converting the
"weed-grown wilderness" into a park, however, presented a daunt-
ing challenge.[17] "Nature," the *Tribune* admitted, "has not been over
lavish to our city in the bestowment of picturesque and romantic
scenery. She will require considerable aid from Art before we can
boast largely of the beauty of the Metropolis of the West."[18] So the
city's next step was to secure a design. Apparently after an abor-
tive competition, the Common Council solicited Swain Nelson for
a plan.[19] Nelson's ongoing work at Graceland, some thirty acres
larger in area than the future Lincoln Park, would by now have
demonstrated his competency for the task and perhaps influenced
the decision to engage him.

The glacially formed terrain of the future park undulated and
swelled in a series of sandy ridges and marshy swales. Nelson's layout is recorded in an elaborate colored plan of the park.[20] (Fig. 4.1)
It shows a nearly bird's-eye view, with trees and shrubs represented
in elevation and landforms rendered with shadows to lend a three-
dimensional effect. The Common Council had mentioned fish ponds
among the possible amenities for the park, and Nelson fashioned a
chain of three linked lagoons and an attendant circular parkland
as the epicenter of the layout.[21] Connected to Lake Michigan by a
stream that bisected the tract, these artificial water bodies emulated
natural lakes in their winding, irregular outlines. Nelson organized
the peripheral land with a labyrinthine drive and walk circuit, assign-
ing carriages to an outer and pedestrians to an inner loop. The drive
and walk layout resembles an octopus, its serpentine trajectories
extending through the projected parkland from the central circle.
He accentuated the landforms with new ridges made from the earth
excavated to form and enclose the lagoons. His walks and drives
presumably followed the contours of the site's rolling topography,

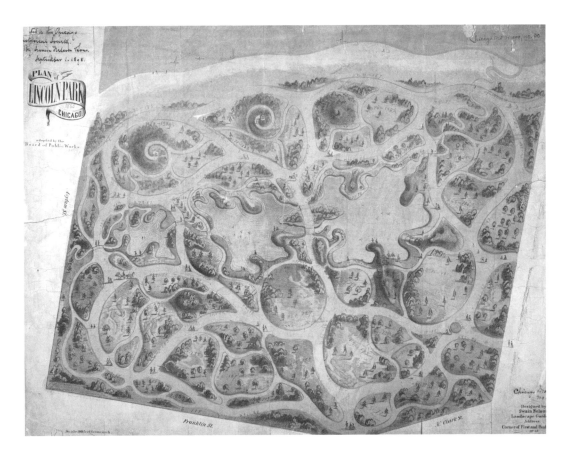

4.1. Swain Nelson's layout for Lincoln Park (c. 1865).
Courtesy Chicago History Museum (ICHi-63843).

both natural and created. Interstitial land was given over to open lawns and partly mantled with irregular masses of trees and shrubs. Accommodating activities such as strolling, riding, fishing, boating, and picnicking, landscape was preeminent in Nelson's vision. Save for eight bridges spanning the lagoons and watercourse, however, his scheme did not include buildings or monuments—a circumstance much changed at today's vastly expanded Lincoln Park. The city contracted Nelson to implement his layout, requiring him to orchestrate earthworks construction.[22] The scope of the project also led him to establish a nursery nearby in order to supply the park with trees and shrubs.[23] He later relocated and expanded his nursery operations in what is now suburban Glenview; he would win acclaim as a nurseryman and later supply plants to Graceland. With the opening of Lincoln Park Chicagoans gained a healthful, bucolic retreat from the city's ever-increasing congestion. By 1867 the *Tribune* could report that many citizens in search of "quiet and pure air" had begun to enjoy the park, "destined to be the most beautiful and popular place of resort in the city."[24]

Nelson's Lincoln Park plan is significant because it is one of the very few known graphic representations of his designs. But it also sheds light on his contributions to Graceland. Lincoln Park was designed some five years after the cemetery and likely represents Nelson's more mature style. A comparison of the park's layout to the initial plan for Graceland shows that although both are curvilinear, Graceland's forms are remarkably simpler, bolder, and less serpentine than Nelson's parkland drive configuration. It is hard to believe that a composition as highly resolved as the cemetery's was Nelson's handiwork, especially given that Graceland was one of his very first projects. Collectively, then, these circumstances suggest that the original Graceland layout was more William Saunders's product than it was Nelson's, and may have been entirely Saunders's.

CONFLICT WITH LAKE VIEW

Graceland continued its aggressive expansion in 1867, when the company made its single largest acquisition up until then, purchasing an additional 109 acres north of Montrose Avenue.[25] Within only seven years, the cemetery had more than doubled in size, compelling its neighbor, the town of Lake View, to action. In 1867 Lake View passed an ordinance to curtail the cemetery's growth—despite the fact that Graceland's state-approved charter permitted it to expand to as much as five hundred acres.[26] That same year the town took a case to the Illinois Supreme Court, attempting to block the formation of a new cemetery within its limits on the grounds it would constitute a public nuisance.[27] The court disagreed, ruling that while a cemetery might indeed "be so placed as to be injurious to public health, and therefore a nuisance," yet "it may, on the other hand, be so located and arranged, so planted with trees and flowering shrubs, intersected with drives and walks, and decorated with monumental marbles, as to be not less beautiful than a public landscape garden, and as free from all reasonable objection."[28] Despite this setback, Lake View's antagonism would prove undiminished. Apart from real or imagined health concerns, there was a more practical reason for the town's continuing opposition: both Graceland and Rosehill were exempt from property taxes. As these and other cemeteries expanded, Lake View's potential tax base contracted, impeding the growing town's development.

In 1868 Graceland's surveyor Samuel S. Greeley completed a

plat to subdivide the new addition into twenty-two new burial sections. Near the entrance at Green Bay Road, Greeley organized the tract with a serpentine road layout evocative of Saunders's original plan.[29] Perhaps owing to ongoing local resistance, however, this plan would not be implemented, and controversy and legal action between Lake View and Graceland would continue for the next eleven years.

H. W. S. CLEVELAND

In 1869 the *Chicago Tribune* twice visited Graceland. In its first report, published in May, an unidentified journalist called the view of the recently expanded cemetery from the main gate "one of the most lovely that can be imagined," adding that it "far transcends any view afforded of Rose Hill."[30] In a city-wide cemetery survey the next month, the *Tribune* called Graceland the "most improved."[31] By this time the cemetery owned 275 acres, of which 86 were improved and in use for burials.[32] Earlier that year Lake View, still an uneasy witness to Graceland's continuing growth, passed another ordinance restricting cemetery expansion.[33] Undeterred, Graceland apparently began contemplating converting more of its undeveloped land to burial sites, which would require significant design work.

Despite Swain Nelson's increased experience at Graceland and his work at Lincoln Park, the cemetery would again engage another designer. Bryan may have considered returning to William Saunders (the two exchanged letters in 1870), but Saunders, now employed by the Department of Agriculture, was at work cultivating Washington's parklands and other civic spaces and likely was not available.[34] Fortunately for Graceland, Chicago had recently become home to another landscape gardener of national standing.

In March 1869 Horace William Shaler Cleveland (1814–1900) moved to Chicago from the East to establish a new practice, opening an office in the same building where, coincidentally, Samuel S. Greeley maintained one.[35] Just before departing for Chicago, Cleveland had been at work for Frederick Law Olmsted and Calvert Vaux; his supervision of plantings at Prospect Park in Brooklyn, New York, was most prominent among his projects with the firm. He was also experienced in cemetery layout. Together with Robert Morris Copeland (1830–1874), with whom he was in partnership in Boston in the 1850s, he designed at least four cemeteries

4.2. Sleepy Hollow cemetery, Concord, Massachusetts. Photograph by Carol Betsch, 2010.

in New England, most notably Sleepy Hollow in Concord, Massachusetts (1855).[36] (Fig. 4.2) There, as Daniel J. Nadenicek and Lance M. Neckar observe, the two landscape gardeners "carefully fitted" the cemetery into a "natural amphitheater" and, as Ralph Waldo Emerson noted in his address at the cemetery's consecration, orchestrated "a 'picturesque effect of familiar shrubs.'"[37] Again quoting Emerson's address, they write that the "'lay and the look of the land' suggested the design" and that Cleveland and Copeland employed art only to "bring out the site's 'natural advantages.'"[38] Once on the prairies, however, Cleveland would discover natural amphitheaters to be in short supply.

Cleveland's presence in Chicago soon attracted local publications. *The American Builder and Journal of Art*, established only months before his arrival, would soon become one of the landscape gardener's most important conduits for publishing and self-promotion. Founded by Charles D. Lakey in October 1868, *American Builder* signaled the growth of the city's fledging arts culture, which Lakey thought substantial enough to sustain a monthly devoted to the subject.[39]

The magazine's first article on landscape gardening appeared in the March 1869 issue, a survey of the ongoing improvements made at Brooklyn's Prospect Park.[40] The next month Dr. John H. Rauch, the champion of rural cemeteries, contributed an article campaigning for public parks.[41] And in May the magazine featured an unsigned article titled "The Architect and Landscape Gardener," which cautioned Chicagoans that "most of our so called landscape gardeners are not educated artists, but simply gardeners; who may be entirely capable of undertaking the nice finishing of garden beds, the shaping of grounds in fanciful forms, and the proper culture of trees, shrubs and flowers: and yet have no such conception of the higher objects of the art, as would enable them to appreciate the general character of a place, and adapt their work to the development of its prevailing expression."[42] One wonders whether the piece was penned by Cleveland himself, for elsewhere in the issue *American Builder* announced it was "pleased, a few days since," to have met Cleveland, mentioning his long connection with Olmsted and Vaux's firm. Citing New York's Central Park as evidence that the partners were "among the most prominent landscape architects in the country," the magazine predicted that "it will certainly result in profit to Chicago, the accession of such a landscape gardener as Mr. Cleveland, one so honourably connected with the above firm."[43] (Fig. 4.3)

That same month *American Builder* published an essay by Cleveland as a stand-alone booklet, *The Public Grounds of Chicago: How to Give Them Character and Expression.* It was as much an advertisement of his own design expertise as it was an instruction manual. Presumably written to attract work from one of Chicago's newly established park districts (which Cleveland eventually did), it marked the beginning of a campaign of self-promotion. Until the end of that year he would contribute an article to each issue of *American Builder*, on a variety of landscape topics.

Most notably, in October 1869 the magazine published a piece by Cleveland titled "A Few Hints on the Arrangement of Cemeteries." If Thomas Bryan missed it, he would have had another opportunity to read it when it was reprinted in the *Chicago Tribune* later that month.[44] One of the issues Cleveland discussed was certain to be of interest to a developing cemetery: the situation that resulted from the "natural desire of every lot-owner to adorn his own lot by planting whatever trees or shrubs he may happen to

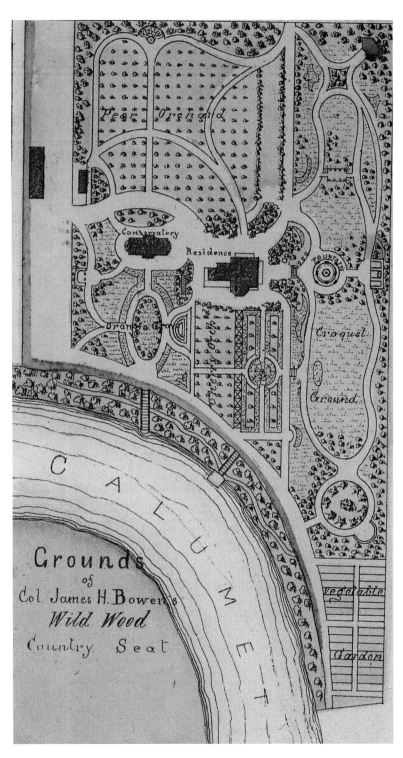

4.3. H. W. S. Cleveland's plan for James Bowen's estate, Wildwood, south of Chicago (c. 1872), one of the landscape gardener's many Chicago commissions.

Courtesy Chicago History Museum (ICHi-29590).

fancy." If individual owners were allowed to fulfill their botanical desires, then the outcome, he feared, would be an uneven mélange of vegetation, and, worse, the maturing trees would "become so crowded as to destroy their individual beauty, and give a sombre and depressing character to the place." Even Mount Auburn cemetery, he wrote, suffered from this "evil." There, when attempting "judicious thinning," cemetery authorities "found that scarcely a tree could be touched without trenching upon the feelings of those who had planted it as a memento." That Cleveland viewed Mount Auburn's planting as defective suggests that cemetery design had entered a new phase since the founding of the pioneering rural cemetery nearly forty years earlier. Rather than horticultural eclecticism, Cleveland urged that the "best development" of a cemetery was one that produced an overall, "general landscape effect"—a holistic planting scheme not for individual plots but for the entire site.

Cleveland's landscape effects took the natural world as their model and required only three main ingredients: "wood, lawn and shrubbery." Wood included both deciduous and evergreen trees; lawn areas could also include, along with turf, "flowers and low shrubs," which "would not interfere with the general character of an open space." The best landscape scenery effects, Cleveland argued, could not be secured "without contrasts of light and shade, which are produced by the proper mingling of wood and lawn"; the beauty of a lawn is developed "only by contrast with surrounding wood." At the same time, he noted, "a continuous wood cuts off the opportunities for extended views, which comprise the chief charm of a landscape, while the true beauty of a wood can only be secured by viewing it across an open lawn." Achieving the aesthetic effects of open lawn would be one of Cleveland's major contributions to Graceland. He strongly advocated barring the practice of "enclosing lots with railings or walls of iron and stone," which "fritters away all possibility of general landscape effect, by obtruding upon the sight in every direction, an endless succession of square pens, enclosed with fences of every imaginable form and color."

Cleveland's hints did not extend to the overall arrangement of cemeteries. Although he urges that they be "properly laid out" and advises designers to record "the location of every avenue and path" on their plans, he offers no clues as to how cemeteries should actually be configured. His mandate to study carefully "the natural character of the ground" in order to identify "the best positions for planting" suggests that this new arrival had yet to encounter the level, low-lying, and swampy ground that typified much of Chicago, including portions of Graceland. Cleveland apparently took fluid, curvilinear drive and path layouts and picturesque, rolling terrain as givens. As we'll see, however, he was not the first to promote these cemetery design tenets.

CLEVELAND AND THE NEW EXTENSION

The exact date that Graceland hired Cleveland is unclear, but we do know that he had completed his plan for the cemetery's improvement by the spring of 1870. It seems likely, too, that the new extension was the product of multiple hands, including the civil engineer William M. R. French (1843–1914), who was professionally connected with Cleveland by July 1870, and the surveyor Samuel Greeley. As *American Builder* announced that month, French had left Chicago's Board of Public Works and was "now connected with Mr. Cleveland in the carrying out of his designs."[45] And in March 1870 Cleveland and Greeley collectively promoted their services in newspaper advertisements.[46]

None of Cleveland's drawings escaped the Great Fire of 1871, so we must rely largely on textual sources to recover evidence of his design work. By the time the cemetery published a new *Catalogue of the Graceland Cemetery Lot Owners* in April 1870, he had finished his plan. In a bid to attract lot buyers, the catalog rhapsodized about the cemetery's design and emphasized Cleveland's new contributions. In a section titled "The Improvement of Lots," the text first distinguishes between Graceland's older and newer sections; the older are apparently the ones Nelson and Saunders had laid out a decade earlier, and they are identified as those where "stone coping is optional, though no fencing can be allowed."[47] The new sections followed a "different plan," one made to "secure a better general effect, and at the same time to reduce the cost to the owners." The next line presumably refers to both the older and newer areas: "These sections were laid out and platted according to the plans of Mr. Swain Nelson." Thus the passage implies that Nelson was the sole author of the cemetery's original and newly extended drive and path layout, which, in turn, fixed the outlines of the burial sections. It continues, "A design for the improvement of the same [sections] has been made by Mr. H. W. S. Cleveland, whose experience and reputation in the decoration of cemeteries afford a guaranty of success." This sequence suggests that the two designers had worked independently and at different times, rather than in collaboration. Curiously, the catalog does not acknowledge Saunders's seminal role or identify what Cleveland was asked to correct. Indeed, until this juncture in the text, the nature of the cemetery's new improvements remains ambiguous. The next sentence, however, clarifies that new "planting will be done by the Company in accordance with this [i.e., Cleveland's] general design in order to secure the best development of the grounds and the most varied effects of light and shade." There is no explicit reference to Cleveland's making any contribution beyond planting design. The design overview concludes with an extended extract—nearly two full pages in length—from Cleveland's *American Builder* article "A Few Hints on the Arrangement of Cemeteries," which, as we've seen, primarily concerned ornamental planting.

The focus on the horticultural dimension of the new plan (together with the dearth of evidence confirming any earlier, large-scale plantings of consequence) suggests that Cleveland was now to orchestrate what amounted to Graceland's first comprehensive planting regime, probably his most significant contribution to the cemetery's development. Engineering activities such as earthworks apparently were not his strong suit, and Greeley and French would have provided those skills. Cleveland's principal talent lay in his masterful ability to make "artistic" botanical compositions. William French, for instance, recalling a sketching trip he made to Chicago's South Park in the 1880s, later wrote: "Across the great lawn, I saw an interesting mass of trees or forest, irregular in outline, retreating at bays, advancing in capes upon the lawn, broken by one or two Lombardy poplars, and I made a rude sketch of it, never suspecting that it was not an accidental effect. When I showed my sketch to Mr. Cleveland, 'Oh, yes,' he said, 'I arranged those trees,' the basis being the fine range of old oaks that ran along a slight ridge in the park."[48]

Although Cleveland's own documentation perished in the Great Fire, one image survives to complement the textual evidence of his design. In March 1878, about eight years after Cleveland completed his work at Graceland and just before William Le Baron Jenney's began in earnest, Charles Rascher (c. 1844–1900), a surveyor who had recently emigrated from Germany, approached cemetery officials with an offer to prepare "a plan of Graceland, showing all the leading monuments with the names of the owners."[49] The outcome of his enterprise is the second oldest known plan of Graceland (after Charles Shober's lithograph) and the only one scholars believe illustrates Cleveland's design contributions.[50] This color map of the cemetery (see fig. 3.3) is a quasi-panorama; like an Egyptian tomb painting, it depicts the ground surface in plan view and renders trees, shrubs, and monuments in elevation. Rascher embellished elevations with shadows, lending an effect of perspective. To the same end, he represented the entry gateway building and the picket fence along Green Bay Road in axonometric projection, a technique in which the horizontal and vertical axes are drawn to scale but the diagonals are distorted, again giving the impression of depth. He also delineated individual burial plots and labeled them with their owners' names. Indeed, it must be noted at the outset that Rascher designed the map to give prominence to the cemetery's monuments, perhaps intending to sell copies to the families that erected them, and that for Graceland itself the map served, at least in part, as a piece of advertising. Hence it is likely that Rascher omitted plantings where they would have obscured the monuments.[51] Nevertheless, his map is still valuable evidence of the cemetery's layout at this time.

Comparing Rascher's map to Shober's lithograph of around 1860 reveals that the cemetery had gained eleven new sections. Continuing Saunders's alphabetic scheme, these are labeled M through W. In Rascher's image, Sections M through R fill the undeveloped northern land. His map also records new burial plots in the northeast part of the cemetery, in Sections T through W. Collectively, these additions extended the area available for burials to Montrose Avenue. Another addition, S, appears near the southwest corner. This one, however, appears to be a replatting of an area first laid out by Saunders. The outlines of the new sections may well have been made by Cleveland. If, however, as the *Catalogue of the Graceland Cemetery Lot Owners* implies, Swain Nelson authored the new and extended drive layout, then he must have taken his formal cues from Saunders. Unlike Nelson's serpentine Lincoln Park road layout of five years earlier, Graceland's new sections are comparatively fluid, seamless extensions of Saunders's original plat.

The map's pictorialized style, somewhat naïve or archaic to our eyes today, does not help us discern the intricacies of Cleveland's planting compositions. It is impossible, for instance, to distinguish his plantings from either earlier plantings or the preexisting oak groves. On the map, massed tree and shrub planting is essentially confined within four areas. The first is Saunders's grotto, which appeared in Shober's lithograph only in outline. Now it is shown as mantled with lawn and evergreen trees, which could have been added by either Saunders or Cleveland. The second is nearby Section K, triangular in shape with concave sides and a curved tip; it is accentuated at the points with tree plantings. This particular section, a focal point within Saunders's original layout, resembles a small park more than it does a burial site. (Graceland founders William B. Ogden and Edwin H. Sheldon, among others, are buried there.) The third locus of substantial planting is strip of vacant land abutting a segment of the cemetery's eastern boundary, and the last is a tract at the northwest corner, which the cemetery did not own at the time Cleveland worked there.[52] Thus, at least as represented on this map, isolated specimens and lawn predominated at Graceland, rather than concentrated groupings of trees and shrubs, although again this may be a consequence of the decision to foreground the monuments.

Rascher also portrayed a body of water, labeled "The Lake," at the cemetery's northeast corner. We've seen that Shober's lithograph included an unidentified water body in approximately the same position, its nebulous form suggesting it might have been more a swamp than a lake. On Rascher's map it has gained not only a name but emphatically revised margins and a trio of islands, planted and platted as burial plots. Was this reconfiguration the product of Cleveland's hand? One period text implies that his layout did indeed feature "miniature lakes," but the circumstances surrounding the map's origins more strongly suggest that the cartographer's ornamented lake was a conjectural representation of an anticipated improvement.[53] Graceland would not gain a bucolic water feature until long after Cleveland's departure from the project.

Although Cleveland himself did not identify the design influences he brought to bear on Graceland, his inspirational source appears to have been Spring Grove Cemetery in Cincinnati, chartered in 1845. (Fig. 4.4) In a short text published in 1872 in connection with the Lake View controversy, Graceland's secretary, William C. Reynolds—most likely relying on information he received from Bryan—noted that the cemetery's "newer sections" had been modeled after "the plan first adopted at 'Spring Grove.'"[54] He was referring to what was known as the "lawn plan," originated by Adolph Strauch (1822–1883) at Spring Grove.[55] Following Strauch's ideals, "stone coping, hedges and side-paths" were to be prohibited at Graceland. The cemetery's "entire planting," Reynolds wrote, was "done under the direction of an accomplished landscape architect," and as a result "each section resembles a beautiful lawn, covered with green turf, and dotted with graceful trees." Trees for these "artistically arranged" grounds had been selected and located with "due regard . . . to their several characteristics, so as to insure varied alterations of light and shade, a harmonious growth, and permanent beauty."[56]

4.4. Spring Grove Cemetery, Cincinnati, Ohio. Around the 1870s, Spring Grove supplanted Laurel Hill as Graceland's main design inspiration source. From the Robert N. Dennis Collection of Stereoscopic Views, New York Public Library.

The lawn plan model, as Walter L. Creese has succinctly put it, "was of a continuous green lawn passing beneath a canopy of trees."[57] Noël Dorsey Vernon writes:

> Strauch's "lawn plan" for Spring Grove . . . was revolutionary in that it established a unified picturesque landscape in which a few fine stone monuments and sculptures, framed by trees, would provide memorials to the dead. In his plan, traditional headstones could not exceed a height of two feet except "extra fine works of art, and by special permission from the Board." The cemetery designer would determine all site grades and create an overall planting effect. Private enclosures and plantings were discouraged. This visual unification of the landscape had been lacking in earlier "garden cemeteries."[58]

By the 1860s Spring Grove had expanded to 412 acres and won wide acclaim. Strauch's success "led to requests for advice and assistance with cemetery design and park projects elsewhere."[59] One such appeal emanated from Chicago, and in 1864 Strauch traveled there to lay out Oak Woods Cemetery at the city's south edge. Through Strauch's guidance the 180-acre cemetery soon took on a "park-like appearance."[60] A late nineteenth-century source described Oak Woods as "conducted entirely upon the lawn plan," with "splendid vistas of waterscape from different points about the four ornamental lakes."[61]

At Graceland, Cleveland suffused Saunders's original rural cemetery layout, by now stylistically archaic, with Strauch's novel "lawn plan" ideals, and the Cincinnati cemetery now supplanted Laurel Hill as Graceland's touchstone.[62] As Creese observes, the "initial rural cemetery type of Mount Auburn was being gradually readapted through the lawn system of Spring Grove into the greater lateral spread of Graceland, harking back to the Grand Prairie" of Illinois.[63] This readaptation would continue in the coming years.

THE LAKE VIEW CONFLICT CONTINUES

Shortly after Cleveland finished his plan, James B. Waller, the very man who had sent Swain Nelson to spy out Bryan's plans for a new cemetery a decade before, opened a new episode in Lake View's continuing resistance to Graceland's expansion. In March 1871 he made a public address to the town's board of trustees, subsequently published under the title *Right of Eminent Domain and Police Power of the State: Trustees of Lake View Township against Graceland Cemetery Company.* Despite the town's defeat four years earlier in the Illinois Supreme Court, Waller's opposition to the cemetery was undiminished. Progress, he believed, was an "inexorable force from destiny" that would overrun and subsume cemeteries; he argued that "providence indicates the development and growth of Chicago to an extent in the future which will demand as a necessity the *ultimate removal of Graceland Cemetery.*"[64] As Creese writes, "The message was unmistakable"—"Destiny would push memory aside."[65]

Graceland's secretary, William C. Reynolds, responded in September, later publishing his rebuttal as *The Limit of the Police Power in the Control of Corporations: A Statement of the Condition, Property, and Franchises, of the Graceland Cemetery Co., together with a Review of the Legislation concerning Cemeteries in Lake View.* Waller had argued that the "continued presence" of the dead impinged on "the enjoyments of the living" and that close proximity to cemeteries diminished real estate values.[66] Reynolds, likely again Thomas Bryan's voicebox, replied that Waller would not "find either his comfort decreased or his property lessened in value" once Graceland's anticipated improvements were completed: "Many of the most costly and elegant residences near Boston, Washington, and Cincinnati, are erected in close contiguity to cemeteries, and their proprietors certainly do not find in the romantic glades and shady dells of 'Mount Auburn,' 'Oak Hill,' and 'Spring Grove' aught to offend the taste or the imagination. But Mr. Waller stands affrighted by some doleful spectre, the creation of his own exuberant fancy; and he brings to the discussion the zeal of an adversary as well as the skill of a lawyer."[67]

Reynolds did concede that some of Graceland's land remained unimproved, "low and unsightly."[68] But had "they not been purchased by our company, it is likely that they would never be devoted to anything more picturesque than a cabbage garden." When the unimproved areas are "drained and planted, and the wild grass gives place to green sod and flowering shrub; when miniature lakes and winding paths shall replace the swale, and monumental marbles relieve by their whiteness the deep green of the trees which

4.5. Chicagoans fleeing the Great Fire of 1871. At left is the City Cemetery. Courtesy Chicago History Museum (ICHi-02881).

surround them, can it be supposed that any one will then wish the landscape changed back to its state of nature?"

There is no known record of James Waller's reply to Reynolds, but in any event disaster would soon intervene and defer the contest.[69] In October 1871, what quickly became known as the Great Fire consumed some two thousand acres of Chicago. (Fig. 4.5) No longer was it the Garden City. Up until then, as one resident later recalled, the "noble, lake-bordered expanse" was "divided into lordly domains, embellished with lovely gardens." Not only was "every street shaded, but entire wooded squares contain[ed] each only a single habitation, usually near its centre, thus enabling their fortunate owners to live in park-like surroundings."[70] All of that was swept away by the fire that began in the evening of October 8. Thomas Bryan opened Bird's Nest to fire refugees and provided them with shelter, food, and clothing.[71] He also personally sustained tremendous financial losses, including the destruction of Bryan Hall and the Graceland offices within, as well as the majority of the cemetery's records.[72] Cleveland's office and papers, along with Samuel S. Greeley's, met the same fate. Matters were soon made worse by the Panic of 1873 and the economic depression that followed it. For the next several years there would be no funding for new landscape projects at Graceland Cemetery.

These economic difficulties may also have interrupted or delayed the implementation of Cleveland's new layout. We do know that it had started or was resumed by 1872, for that December, city surveyor George Frost made a survey plat of Section M, the first of the cemetery's new burial sections.[73] If Graceland had begun laying out the new ones in alphabetical order, then December 1872 might possibly have been the start date for implementing Cleveland's scheme.

Lake View's opposition to cemetery expansion was unrelenting. In September 1873 the town returned to the Illinois Supreme Court, this time in a case against Rosehill. Upholding the cemetery's charter, the court ruled in Rosehill's favor. "The sentiments of our better natures and the civilization of the age," the majority opinion asserted, "demand that these sacred places shall be made attractive and beautiful by the employment of the highest skill in landscape culture"; such cemeteries were "very far from being a nuisance" and instead "attract hither, as to pleasant places, lovers of the beautiful in nature, as to groves and parks that have been adorned by the lavish expenditure of money."[74] Rosehill's victory was a narrow one, however; three of the seven judges dissented. Lake View Township, its geology and soils ideal for burials, was now home to eight cemeteries. Were these to expand, the dissenting justices feared, overcrowding would result. Echoing James Waller's "city of destiny" sentiments, one justice asked, "Is it unreasonable to anticipate, in view of the marvellous growth of the city of Chicago, that, at no far distant day, the ground will be found to be in the midst of the dense population of a city?"[75] The close decision may have offered Lake View a modicum of consolation, and the town's challenges would increasingly worry Graceland. And, although its holdings were still not in excess of the five hundred acres its charter allowed, the next year Thomas Bryan and the board of managers apparently decided that the cemetery already had land enough. In September 1874 the board resolved that "Graceland Cemetery shall not be extended to include any part of the Company's land" lying "west of the Green

Bay Road" or "in the North half of Section 17. 40. 14" or "north of Sulzer Street," nor would it expand eastward "beyond the East line of the West quarter of the South East quarter of Section Seventeen."[76] Instead, these properties would be held "or disposed of for other than burial purposes." The company would later commission a landscape gardener—O. C. Simonds—to lay out some of these now extraneous holdings as residential subdivisions prior to their sale.

FIVE

Bryan Lathrop and
William Le Baron Jenney

In 1877 Thomas Bryan, a consummate politician who was also
renowned for his civic stewardship, was called to take a post in
the Rutherford Hayes administration as one of the three commissioners of the District of Columbia.[1] Just before his nearly twenty-year
tenure with Graceland came to an end, he hired the architect and
landscape gardener William Le Baron Jenney (1832–1907) for what
would prove to be one of the cemetery's last major landscape initiatives. (Fig. 5.1)

In the spring of 1877, as the financial effects of the Panic of
1873 began to abate, the cemetery's board of managers decided
to begin draining Graceland's undeveloped land, low-lying and to
the east, in order to create new burial sites.[2] They were perhaps
anticipating a ruling by the Illinois Supreme Court that Graceland's properties not yet in cemetery use would be subject to taxa-
tion; no doubt in part they sought to avoid any chance of having
to pay taxes on this land to the town of Lake View.[3] The expansion was announced in a new promotional campaign. In April
Graceland ran two different display advertisements in the *Chicago
Tribune*. On April 10, in a short ad ostensibly attempting to dis-
pel the "erroneous impression" that the "best portions of Graceland" had already been sold, the company announced that it was
offering "for the first time, some of the most attractive sections
of its grounds" and that its managers had "prepared plans for a
comprehensive system of improvements, and purpose to maintain Graceland in the front rank of American Cemeteries."[4] Five

5.1. William Le Baron Jenney (1832–1907).
Courtesy Chicago History Museum (ICHi-19760).

days later the company placed a second advertisement that would run six more times during the next few months. This ad extolled the cemetery's advantages: "great natural beauty," "undulating surface and fine trees," and "a gravelly subsoil, giving perfect underdrainage," attributes Graceland had promoted ever since its dedication nearly twenty years earlier. But it also added that "living springs of water" would "supply the lakes which are to be made and which have been begun." After noting that the company owned "about 200 acres of land besides that already subdivided," the ad promised that "newer sections, and all those which shall be added in the future, will be maintained on the 'lawn plan.'"[5] At a meeting in June the board of managers voted unanimously to construct a "ditch and tile drain" from the "low ground on the eastern side of Graceland Cemetery, along Sulzer Street to Lake Michigan"—the cost to the company not to exceed fifteen hundred dollars.[6] Indeed, the board was ever mindful that transforming the sloughs would require "large and costly improvements" before the area could "by the art of the architect and landscape gardener be made healthful and beautiful."[7] Making the tile drain, along with the design and construction of artificial lakes and laying out new burial sections resonant with Strauch's lawn plan ideals, were tasks requiring a professional landscape gardener, ideally one skilled in drainage techniques.

WILLIAM LE BARON JENNEY

Although he is most often remembered today as the architect of the first steel-framed skyscraper, the Home Insurance Building (1885) in Chicago, in Thomas Bryan's day William Le Baron Jenney was known as a landscape gardener as much as an architect. Indeed, in nineteenth-century Chicago little distinction was made between the two pursuits, so long as the talent and will were present, and it was not anomalous that he won the Graceland commission. And, as Walter L. Creese notes, Jenney was "not only the cool technician-engineer he is often depicted [as], but also, like Olmsted, an environmental strategist."[8]

What brought him to Bryan's notice? For one thing, Bryan was probably already aware that Jenney had been substantially involved in two significant projects in the Chicago area—the newly constructed town of Riverside, and the improvement of Chicago's western parklands. He may have known too that Jenney had designed a cemetery in Moline, Illinois, also named Riverside.

In 1869, when H. W. S. Cleveland began publishing essays such as *The Public Grounds of Chicago* to herald his professional arrival, Jenney had already been working in the city for about two years, and a number of notices of his work had appeared in *American Builder and Journal of Art*. In October 1868 the premiere issue featured two plates illustrating his and his partner Sanford Loring's designs for a pair of Chicago houses, which would later appear in their own book, *Principles and Practice of Architecture* (1869). Describing Jenney as "an accomplished architect of this city," the magazine "cordially commended this forthcoming volume to the building public."[9] Loring and Jenney's book advertised the pair's talents as including not only park but also cemetery design. Of no less interest to Bryan, the architects, as Creese notes, claimed to "follow Downing before all others."[10] Writing nearly twenty years after the pioneering landscape gardener's death, the two designers observed, "There is a want of intelligence in matters of art in American country villages, especially in the West; such books as Downing's have done much to supply this want, and should be more generally read."[11] The next year *American Builder* again featured Jenney and his work. In its February 1870 issue, for instance, the magazine reported that he had "recently been elected engineer and architect for the West Chicago Parks" and endorsed him as a "gentleman of fine taste and thoroughly qualified for the position."[12] The following November it illustrated Jenney's plans for a villa to be erected in "elegant" Riverside. Highlighting the architect's landscape gardening skills, the journal observed that "the grounds are designed in keeping with the house."[13] The next month, *American Builder* reviewed Jenney's design for another Riverside dwelling, and added that throughout the town "great care is being taken with the private as well the public grounds, which adds not a little to the general attractions."[14] Jenney was, of course, one of the professionals taking "great care" in orchestrating Riverside's landscape development. In 1871 the magazine featured Graceland surveyor Samuel S. Greeley's article "Laying Out of Western Suburban Villages," in which he noted that Riverside was the "most carefully planned" and "most successfully built" town laid out on a curvilinear plan.[15]

As a young man, Jenney decided to pursue a career in civil engineering. A Massachusetts native, he enrolled in Harvard's fledgling program, but he soon became dissatisfied with it and looked to Paris, then a center of engineering and architectural education. In 1853 he was admitted to the École Centrale des Arts et Manufactures to study engineering and architecture.[16] The new American student was a firsthand witness to Baron Georges-Eugène Haussmann's transformation of the French capital, including the construction of a new park system orchestrated by Jean C. A. Alphand.[17] This experience would help shape Jenney's landscape gardening ideals.

Immediately after graduating in 1856, he sailed to Mexico and began work for the Tehuantepec Railroad Company, then constructing a line across the isthmus. Financial crisis the next year soon ended the project, however, and Jenney returned to Paris, where he spent a few more years in work and study.[18] In 1859 he was back in the United States, working for the Marietta and Cincinnati Railway Company, and by 1860 he was settled in Cincinnati and at work mapping a rail line between that city and nearby Loveland.[19] The new line passed near Spring Grove Cemetery; by 1860 the cemetery would have begun to show evidence of

Adolph Strauch's design contributions, including large-scale earthworks and the construction of water bodies, and Jenney may well have found his way to the place. After his railroad surveys, Jenney opened his own architectural and engineering practice.[20] The Civil War soon intervened, but rather than interrupting his career the war actually provided him with opportunities to extend it: he joined the U.S. Army Corps of Engineers and was soon designing fortifications and bridges for the Union side, ultimately working under General Sherman.[21] He also met Frederick Law Olmsted, then executive head of the United States Sanitary Commission, at the siege of Vicksburg—a meeting that would prove to be significant for his future.

After the war ended Jenney accompanied Sherman to St. Louis, where he spent some time compiling a map recording the general's campaigns. By December 1865 Jenney's professional ambition had expanded to include landscape gardening, and he wrote to Olmsted that month, "There is no situation that I can imagine where I should derive so much pleasure from the work I might be called upon to perform as one in which Architecture, Gardening and Engineering were associated, and I most earnestly desire and hope that some such position be within my reach."[22] This letter, written in the hope of securing employment with Olmsted, was Jenney's second communication with him that month; he had first written when he learned that Olmsted and Vaux were soon to resume their Central Park posts and that they also anticipated the Prospect Park commission. Jenney lacked the financial resources to restart his own practice, and there was another reason as well: "In the West," he lamented, "there [is] little knowledge and little desire for art: besides one here must live within themselves [*sic*] without the means of profiting by the works of others."[23]

Jenney resigned his military commission in May 1866. According to a short biography that appeared in 1892 and was likely based on information supplied by Jenney himself, he "entered the office of Olmsted, Vaux and Withers, of New York city, architects and landscape artists."[24] His tenure with the firm was apparently a brief one; by the fall of 1867 he had relocated to Chicago and entered into a partnership with the architect Sanford E. Loring (1841–ca. 1910). Presumably Chicago's promise of opportunity for financial success was alluring enough to attract Jenney back to the artless West—or perhaps he hoped to fill the aesthetic void. Jenney's practice with Loring lasted only two years, but despite the dissolution of the partnership in 1869, that year would prove a good one for Jenney: Chicago's newly formed West Park district appointed him its "architect and engineer," and the same year he took up a position as the Olmsted firm's "superintendent of architectural construction" at Riverside, where Jenney was to implement Olmsted and Vaux's design. Jenney now found himself in immediate need of "careful and skilful associates," something then still in short supply in Chicago, and consequently he formed a partnership with a trio of former Olmsted associates from New York.[25] This arrangement was apparently conceived as short-term from the outset and lasted only a single year. Along with the local park and suburb work, Jenney and his partners are known to have laid out portions of Washington Park in Albany, New York, and at Nashville, Tennessee, they planned "the improvement of the capitol grounds," presumably a prestigious commission gained through Civil War connections. Unfortunately, we know little more of these latter projects. But Riverside and the West Parks were among the most extensive projects undertaken in Chicago up until that time, and they lie at the heart of Jenney's landscape gardening oeuvre.

BRYAN LATHROP

With the departure of Thomas Bryan to Washington in 1877, his nephew Bryan Lathrop (1844–1916), the son of Jedediah Lathrop, became the vice president of Graceland's board of managers, and he would guide the cemetery's landscape development for nearly forty years. (Fig. 5.2) Lathrop had arrived in Chicago in 1865 after completing four years of travel and study in Germany, France, and Italy, cultivating an appreciation of the fine arts.[26] Visiting museums, Lathrop would "study the pictures, decide in his own mind which were best, and then compare his judgement with that of the best critics."[27] Along with this cultural pursuit, as he later recalled, he visited "the parks and gardens of Europe," gaining "a love for landscape gardening."[28] This was, of course, a passion his father and uncle also shared, as is illustrated by their luxuriant Cottage Hill gardens at Huntington and Bird's Nest. That June, the twenty-one-year-old Lathrop left Virginia for Chicago and joined his uncle's real estate investment practice. Given their mutual landscape gardening avocation, it comes as no surprise that Thomas Bryan soon involved his nephew with Grace-

land's ongoing development. Indeed, the young man would join the board of managers of the Graceland Cemetery Company in 1867.

At the time of Lathrop's arrival, Cottage Hill was a hive of landscape gardening activity. Only a year earlier, for instance, Jedediah Lathrop had begun building Huntington, and now he involved his son in improving its twenty-six-acre grounds.[29] Bryan Lathrop recounted this considerable undertaking decades later. At the outset, the "natural wetness and sourness of the prairie soil" posed a formidable challenge.[30] Jedediah Lathrop remedied these conditions with a "very complete system of underdrainage," one his son believed "more thorough and extensive in fact than had ever been attempted in the vicinity of Chicago." This elaborate subterranean network consisted of "main drain pipes with lateral drain pipes every fifty feet over the entire premises to the south line of the vegetable garden, from four to five feet below the surface of the ground." When Lathrop was later confronted with poorly drained soil at Graceland, similar techniques would be employed. The drainage problem solved, Jedediah next improved the soil quality with "many hundred loads of sand and fine gravel and of horse

5.2. Bryan Lathrop (1844–1916) and his extended family on the grounds of their Elmhurst mansion, Huntington, around 1884. Bryan is standing in the back row, behind his parents, Jedediah and Mariana. Courtesy Thomas Nelson Page Papers, Special Collections Research Center, Swem Library, College of William and Mary.

and cow manure." He was now ready to begin making his garden. Working to the plan of "the best landscape gardener in Chicago" (probably Swain Nelson), Jedediah and his son made drives and pathways "with great thoroughness." The endeavor also entailed extensive planting, including the remarkable transplantation of "a good many elm trees of large size, perhaps a foot in diameter at heavy expense." The estate's improvements, completed in 1866, consumed three years and $65,000, apart from the cost of the land. Graceland superintendent O. C. Simonds, writing many years later, remembered Lathrop as having "helped his father to develop a unique and beautiful home," becoming "a most able amateur in landscape designing" in the process.[31]

At its October 1878 meeting, along with accepting his plans for the cemetery's sewer and expanding the scope of his work, the board of managers instructed Jenney, as the new superintendent, to work "under the directions of Bryan Lathrop."[32] This arrangement, Simonds later wrote, reflected Lathrop's "influence as an amateur in landscape design," and it would be of profound consequence for Graceland's future development.[33] As Swain Nelson's name is absent from the meeting's minutes, he presumably no longer continued as assistant landscape gardener; it seems Lathrop himself had now effectively replaced him, and thus became not only Jenney's client but also his supervisor. Indeed, in the wake of Thomas Bryan's departure, Lathrop, serving as both vice president and secretary, increasingly assumed responsibility for the cemetery's day-to-day operations and soon became Graceland's "ruling spirit."[34]

By 1877, more than a decade after his arrival in Chicago, Lathrop's own views on landscape gardening had no doubt matured. After first gleaning foundational knowledge touring European parks and gardens, and then furthering his avocation by contributing to the ongoing landscape improvements at his father's estate, he had also continued to read voraciously on the topic. A. J. Downing's works of the 1840s and 1850s were fundamental to his study, just as they had been for his uncle.[35] Among others, a more recent title by Downing's protégé Frank J. Scott (1828–1919), *The Art of Beautifying Suburban Home Grounds of Small Extent* (1870), also captured Lathrop's attention.[36] He complemented these domestic texts with works by British landscape writers, such as William Robinson (1838–1935), Humphry Repton (1752–1818), and William Gilpin.[37] Although today remembered more for championing the English flower and "wild" gardens, Robinson also wrote explicitly on cemetery design. For instance, in 1876, around the time Graceland was apparently considering expansion, he praised Spring Grove Cemetery's "plans and principles of management" in an essay on what he called America's "Garden Cemeteries."[38] Believing Spring Grove to have "produced even better results than are seen elsewhere in America," Robinson illustrated the piece with engravings made from photographs supplied to him by Frederick Law Olmsted. In what may well have been an allusion to Graceland itself, he cited Chicago as one of the places where cemeteries "owned by one or more individuals" could be found.[39] Of course, Spring Grove's exemplary status was not news to Lathrop. If he did read Robinson's essay, this endorsement from overseas probably further entrenched his esteem for the place. Most of the sources known to Lathrop advocated a "naturalistic" landscape gardening aesthetic; surveys of regular or formal gardens are notably absent. By the 1870s Gilpin's and Repton's works were many decades old and possibly considered of merely historical interest. If Lathrop had been seeking contemporary design advice, some of the volumes he turned to were arguably archaic; nonetheless, his reading list bespeaks a landscape taste that esteemed picturesque beauty as timeless.

Lathrop would also have had firsthand contact with H. W. S. Cleveland, at least during his Graceland tenure. Already a member of the board of managers in 1870, Lathrop would have participated in the decision to commission Cleveland—if only to endorse what was probably his uncle's choice. One can imagine Lathrop taking every available opportunity to discuss landscape gardening with one of the very few professionals then practicing in Chicago; in fact Simonds attributed Lathrop's knowledge of the subject partly to "his association with Cleveland [and his partner William M. R.] French."[40]

Probably by 1877 Lathrop had become, as he later characterized himself, "as closely in touch with [landscape gardening] as a layman can be."[41] He later set forth his views publicly in a lecture, "Parks and Landscape Gardening." Believing landscape gardening to be "one of the greatest" of the fine arts, he championed the natural world as its most authentic inspirational source and model. By contrast, the "old formal gardens of Italy," for instance, were anathema. For Lathrop, Italian gardens and their contemporary derivatives were "unlovely" exemplars of "stiff formalism," evocative of landscape gardening's "dark ages." Rather, he drew his design ideals from an English font. Outlining a hypothetical tour of European gardens, he sardonically observed: "If, by chance, you come upon a charming bit of turf, with masses of flowering shrubs and trees not in lines and left to grow untrimmed, you are told—it may be in Italian, or German, or Spanish, or French—that this is the 'English Garden,' and you say to yourself, 'God bless it.' There is a touch of nature in it." As he urged landscape gardeners to produce "not copies," but "perfections of nature," Lathrop's ideal garden displayed only a touch of nature. As we will see, Lathrop's ideas were broadly compatible with those of Jenney.

By 1877 Lathrop appears to have not only formulated his own design ideals but also begun applying them directly at Graceland, perhaps even in advance of Jenney's engagement. That year, for example, he advised he cemetery's superintendent, J. S. Birkeland, that he would permit, in one particular instance, stone coping to be placed around a grave, "provided it be sunk so as to admit mowing of the grass over it or about it."[42] Also around this time, Lathrop had concluded that comparatively expansive burial plots were necessary to achieve "artistic landscape effects."[43] At the time, "nearly all the lots in Graceland were of uniform size, 15 by 25 feet," and buyers desiring a large lot had to purchase two or more. So he now proposed to divide "an entire section into large lots, no lot being less than 50 by 60 feet." Simonds later recounted that the board members "shook their heads, but finally consented" to these expansive new dimensions. Section O, initially laid out by Cleveland, became the first to be resubdivided in accordance with Lathrop's idea.[44] It would not be the last. Subdividing a section, however, was not a task equal to converting a swampy expanse, some fifty acres in extent, into a cemetery. That would require the skills of a professional such as Jenney.

JENNEY'S WORK AT RIVERSIDE, THE WEST PARKS, AND MOLINE

By 1868 Frederick Law Olmsted and Calvert Vaux were widely known for projects such as Central and Prospect Parks. Among the Chicagoans attracted to their work was Emery E. Childs, soon to be president of the Riverside Improvement Company, which would be incorporated the next year. Now in need of professional advice and a plan for his suburb building enterprise, Childs commissioned Olmsted and Vaux to design the entire town, which was to be located nine miles and forty minutes from the city's center on the Chicago, Burlington & Quincy Railroad line, at a juncture where it crossed a meander in the Aux Plaines (now Des Plaines) River.[45] There Childs and his company had secured a sixteen-hundred-acre tract. A former farm, much of the property had been under cultivation. Unlike the nearby level grasslands and agricultural fields, here the river's elevated banks offered prospects across the sinuous watercourse, and at its margins the land was mantled with "groves of thrifty and beautiful trees."[46]

Eschewing the street gridiron typical of the region, Olmsted and Vaux composed the new suburb on a curvilinear plan. (Fig. 5.3) The gentle curves of the Des Plaines River, along with Joseph

5.3. Olmsted and Vaux's plan of Riverside, Illinois (1869).
Courtesy Riverside Museum, Riverside, Ill.

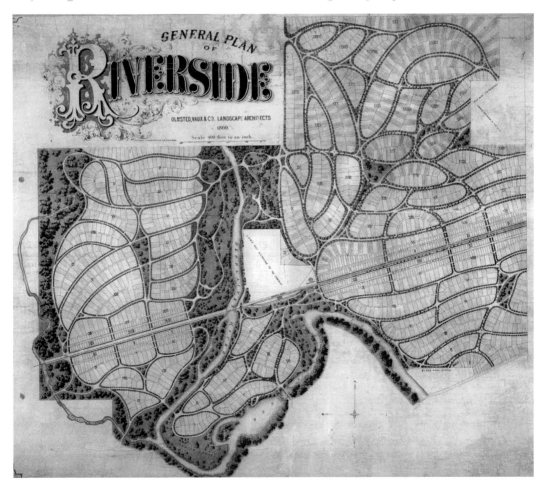

5.4. Scottswood Common, one of Riverside's communal parklands, around 1888.
Courtesy Village of Riverside, Riverside Historical Commission.

Paxton's suburban Birkenhead Park (1847) in Liverpool, England, were likely sources of inspiration for this alternative configuration. Olmsted and Vaux envisioned the suburb as a domestic "bastion against urban congestion," and they reserved nearly half the total acreage for public parklands. One of these, the "Long Common," lies at Riverside's core. (Fig. 5.4) Patterned after New England village greens and accentuated with trees, this rippling swath of turf is effectively the center of the suburb's road network. Hoping to make the parklike atmosphere even more pervasive, Olmsted and Vaux also intended that dwellings be set back thirty feet from the road, expanding the area available for sweeping (and unfenced) open lawns. "Stretched above, but not so high as to challenge the human scale," Walter Creese writes, "was the canopy of green trees sheltering the monochromatic lawn, like a rippling awning, from the infinite and ever-changing prairie sky."[47] In large measure railroads made this bucolic enclave and other early suburbs possible, but Olmsted and Vaux apparently preferred travel by carriage and proposed a tree-lined parkway to link village with city.

After the landscape architects completed their plan in 1869, the task of realizing it fell to their former employees, William Le Baron Jenney and his partners. Also serving as Riverside's architect, Jenney would design an array of houses and public buildings, such as a hotel and a landmark water tower. It is important to remember that Olmsted and Vaux were, after all, based in distant New York and had numerous other commitments, and thus, as Jenney's biographer writes, it would have been difficult, if not impossible, for the two to "oversee every detail of the project."[48] Consequently, "most of the day-to-day decisions such as the placement of paths" and the "grouping of trees," for instance, were now "in the hands of the engineers."[49] The extent of Riverside's projected infrastructure was formidable. It included, for instance, asphalt and stone walks, gravel roads "with solid bases flattened by a fifteen-ton steam roller imported straight from Europe," drain and sewer pipelines, and gas street illumination. These projects were in the charge of Jenney's partner L. Y. Schermerhorn, who, as Creese puts it, "drove hard": "Within two years, nine and a half miles of road, seven miles of walks, six miles of water pipe, five and a half miles of gas pipe with two hundred lamp standards, and sixteen miles of drains and sewers had been put in, together with thousands of trees and shrubs."[50]

Creese reminds us that Chicago in 1870 "had virtually no suburbs," and that the abrupt loss of green space within the city to the Great Fire "explained more clearly the subsequent attraction of it at Riverside." After the fire's devastation, Creese writes, there "was no gradual creep out to the suburb, but rather a great leap from inside the city to the outside."[51] Two of those who made the leap to Riverside were Jenney and Schermerhorn. Both would reside there in houses designed by Jenney.

With the work at Riverside under way, Jenney took on another substantial landscape initiative. In 1869 Chicago established the North, South, and West Parks districts, each headed by an assembly of appointed commissioners. Perhaps owing to their Riverside work and their association with Olmsted and Vaux (themselves retained to lay out Chicago's South Parks), Jenney and his partners won the commission to design the West Parks. In Jenney's vision, according to Reuben Rainey, the three West Parks "would create a 'chain of verdure' and connect the adjacent South and North Park systems."[52] Indeed, Chicago was now resolved to create what no other American city yet had, an interconnected parkland network integrated within its urban fabric. As Theodore Turak points out, even New York's Central Park "remained aesthetically apart"; inserted within the city's street gridiron, Olmsted's park "stands in splendid, rectangular isolation to its environment."[53] In more prac-

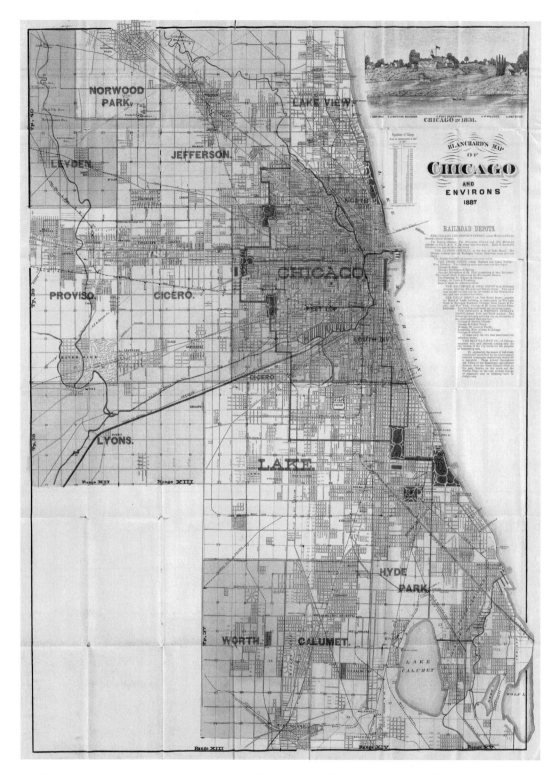

5.5. Chicago's pioneering network of parks and boulevards. *Blanchard's Map of Chicago and Environs,* 1887.

Courtesy David Rumsey Map Collection, www.davidrumsey.com.

tical terms, Jenney's West Parks complex would be anchored by three large parks of roughly equal size: the 171-acre Central (now Garfield) Park, with 171-acre Douglas Park to the south and 193-acre Humboldt Park to the north.[54] (Figs. 5.5 and 5.6) These would be stitched together with eight miles of boulevards at the city's still undeveloped western periphery.

Rainey writes that Jenney's "sophisticated" planting plans structured the parks and "transfigured the 'wild,' treeless prairie into a verdant counterpoint of exotic semi-tropical plants and hardy native vegetation." (Fig. 5.7) Apparently inspired more by the planting strategies employed in the parks of Paris than by Olmsted and Vaux's examples, Jenney's botanical compositions were quite different from "the stock-in-trade 'pastoral-picturesque' planting schemes of many large nineteenth-century American urban parks."[55] As we will see, Jenney's planting scheme for Graceland would not be realized, but there is another aspect of his West Parks designs that would prove of the greatest importance for the cemetery. For Jenney, drainage was one of the paramount "modifying influences" dictating the ultimate form of his park layouts. Turak writes:

> [Jenney] computed the rainfall for the Chicago area and then estimated the loss of moisture through ground absorption, evaporation, and surface runoff. He concluded that the capacity of the existing sewer system was inadequate and that new lines would cost about $200,000. If he employed reservoirs of some sort, the cost to control drainage could be cut to $40,000. Ornamental lakes and rivers solved the problem. Jenney covered an unusually large area of his parks with water. The depth of the water thus stored could be controlled depending upon weather conditions and need. During the skating season, for instance, the depth of any of the lakes could be made less than four feet for safety reasons. Earth removed to form these lakes provided material with which he shaped the landscape.[56]

This was precisely the technique Jenney would soon apply in his new position at Graceland.

Jenney was already experienced in cemetery layout before he took up the Graceland commission. In 1874 he designed a rural

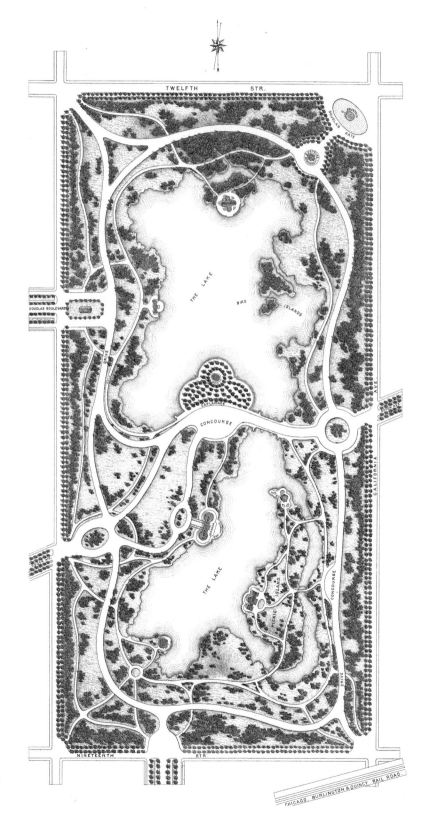

5.6. William Le Baron Jenney's layout for Chicago's Douglas Park (1871). Jenney was then in partnership with engineers Louis Schermerhorn and John Bogart.
Courtesy Chicago Park District Special Collections.

5.7. Humboldt
Park as realized in
Jenney's design.
From Simon, *Chicago, the
Garden City.*

cemetery overlooking the Mississippi River in Moline, a city in northwestern Illinois that was the home of the agricultural equipment company founded by John Deere. Two years earlier, the eminent Moline citizen Charles H. Deere (1837–1907), John Deere's son, had commissioned Jenney to design his new residence, Overlook. The architect responded with a three-story "Swiss villa," reminiscent of his earlier Riverside houses. As its name suggests, the house occupied a commanding hilltop position, with prospects not only to the river but also to the Deere factory.[57]

Deere's new house plans were finished by March 1872. A local newspaper predicted that it would be "one of the finest residences in the county, equipped with all the modern conveniences."[58] Jenney probably also designed Overlook's seven-acre surrounds. Contemporary photographs reveal a winding approach road, terminating in a circular drive and porte-cochere entry. The grounds, resembling more a private parkland than a garden, included an array of outlying buildings, such as a gazebo, a greenhouse, and a carriage house.[59]

In 1873, around the time Overlook was completed, Moline elected Charles's father, John, who had largely retired from the family business, as its mayor.[60] Acting in this new capacity soon

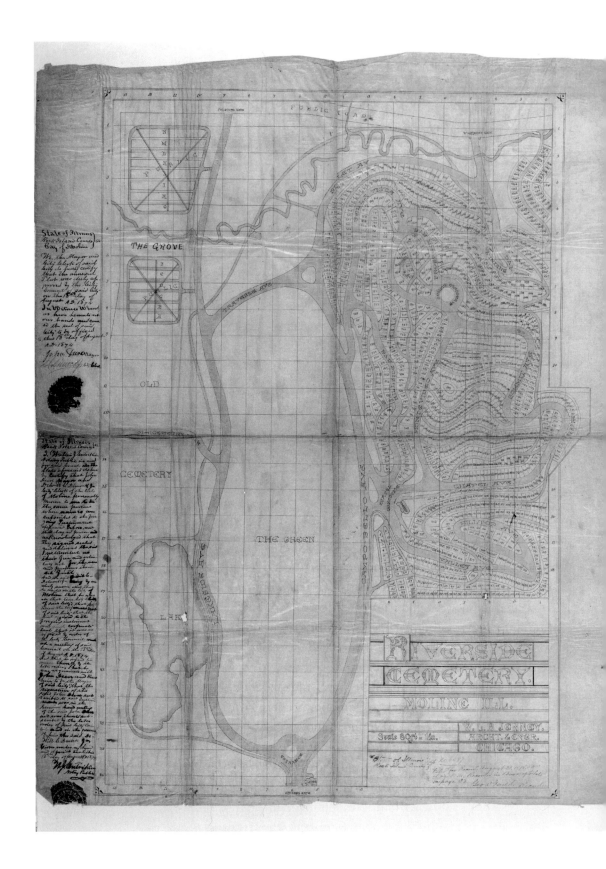

involved him with the town's takeover of the six-acre Moline Cemetery at its rural outskirts, which had been founded in 1851. Deere and the city council appointed a board of directors to administer the cemetery; this new body elected to change the cemetery's name to Riverside Cemetery and to enlarge it, and they soon secured a seventy-four-acre addition.[61] Now a landscape gardener would be needed for a plan, and in February 1874 the board, most likely on the recommendation of either Charles or John Deere, awarded Jenney the commission to lay out the new extension.[62]

Jenney recorded his design in a remarkably ornate colored plan, one of the very few of his landscape design plans to survive.[63] (Fig. 5.8) He organized Riverside Cemetery with a serpentine carriage drive and pathway network. As his avenue names—Summit, Hilltop, Prospect, Valley, and Hillside—indicate, the cemetery occupied a rolling, irregular terrain, and much of the sinuous drive followed the land's contours. He refashioned the peripheral slopes in a series of terraces, subdividing them into fourteen hundred burial plots.[64] Yet Jenney's curvilinear layout was not entirely geologically determined. He transformed the open ridgetop into a parkland, shown on the map as "The Green," to buffer the cemetery's old section from the new one. Unlike the steep slopes at the site's margins, here at the summit the ground gently undulates. This surface would have permitted a regular or geometric composition, had Jenney wanted to make one. Instead he opted to incorporate "The Green" seamlessly into his wider, fluid layout. Though Jenney's drawing does not represent vegetation, from the character of the roads and terraces we can infer that Riverside's planting configurations would have been picturesque. Manipulating an existing watercourse, Jenney ornamented the grounds with a spring-fed lagoon, labeled "Lake," with an irregular outline reminiscent of those he designed for Chicago's West Parks. There was also a naturally occurring water feature, Chub Creek; this stream slices through a portion of the cemetery's eastern extreme, its presence registered by the name of the drive that spanned it, Rivulet. The names Jenney gave to the roads, themselves evocative of a Grand Tour itinerary—Bella Vista, Campo Bello, and Montevideo—suggest that the elevated, panoramic views from the cemetery must have been spectacular.

By 1876 the directors had spent almost $5,000 implementing Jenney's design, and the city directory promoted the cemetery as "a

(*Opposite*) **5.8. Jenney's layout of Riverside Cemetery in Moline, Illinois (1874).** City of Moline, Parks and Recreation Department.

spot favored by nature and beautified by art."[65] It became the final resting place of both John and Charles Deere.

JENNEY'S WORK AT GRACELAND BEGINS

Jenney's association with Graceland had a seemingly inauspicious start. In October 1878, one of his first documented jobs for the cemetery was to make not a landscape composition, but a brick sewer.[66] In the words of the company's secretary, William Reynolds, the tract to be developed was "swaily" or "a slough," and it would require "a large amount of money for improvement, until a sewer should be built to the lake."[67] Thus Jenney's expertise in drainage may have been fundamental to his winning the Graceland commission. Accepting his completed design and specifications for the sewer at its October meeting, the board of managers next ordered construction of a pipeline.[68] At the same time, it also dramatically enlarged the scope of Jenney's work. He was now to be "Superintendent" of not only the cemetery's drainage and engineering, but also its landscape gardening.[69]

In the autumn of 1878 Jenney assigned management of the sewer project to his new assistant, Ossian Cole Simonds (1855–1931).[70] A native of Grand Rapids, Simonds had graduated that year from the University of Michigan, where he had studied civil engineering and architecture, partly under Jenney's instruction.[71] Having come to Chicago only months earlier to pursue a career in architecture, Simonds, like his employer, was also to practice as a civil engineer and, soon, as a landscape gardener. He later described the site of their project. Unlike the oak-studded, rolling lands within the cemetery's original plat, the area now to be developed was "treeless" and "low," its porous terrain a mosaic of "swamp," "slough," and "celery field."[72] The first task at hand was to drain a slough at the cemetery's northeast. As we know, Charles Shober first recorded this naturally occurring landscape feature in his lithograph of the cemetery's original plat around 1860, and Charles Rascher later idealized it as "The Lake" in his Graceland map of around 1878. Jenney, using data collected by Simonds after "run[ning] a line of levels half a mile north to Lawrence Avenue and then half a mile east to Lake Michigan," determined the water level "could be lowered five feet."[73] To achieve this, he directed the construction of "a thirty-inch brick sewer from Lake

Michigan west to what is now Broadway" (then Evanston Road) and a "vitrified pipe line" linking the "sewer to the lagoon to be drained."[74] With the creation of the subterranean infrastructure now under way at the cemetery, the scope of Jenney's work would expand the following spring.

SIX

Final Expansion

In April 1879, after more than a decade of conflict, Bryan Lathrop at last resolved Graceland's long-standing dispute with Lake View. In January, the cemetery's decision to convert about 190 acres of its undeveloped land to burial sites had triggered a new episode in the adjoining town's continuing opposition.[1] The next month Lake View amended its charter to "forbid the use, save with the Town's consent, for Cemetery purposes, of grounds not already enclosed and platted for such uses."[2] As we have seen, the town had similarly amended its charter in 1867 and 1869. Graceland, relying on the state-sanctioned charter that permitted it to expand to up to five hundred acres, now apparently threatened to challenge the town in the Illinois Supreme Court.[3] Such legal action was likely to be protracted and costly, of course, and no less troubling for Graceland was that Lake View had very nearly won its 1873 Supreme Court case to block the expansion of Rosehill Cemetery.

Breaking the stalemate with the town later that month, Lathrop initiated a series of conferences with a committee of "citizens taking a special interest in the matter," including Graceland's original opponent, the attorney James B. Waller.[4] A compromise resulted in March.[5] (Fig. 6.1) If the town permitted "some 35 acres, partly already owned and partly to be acquired" to be used for burials, Lathrop would agree to accept the extended boundaries "as the fixed limits of Graceland Cemetery, and relinquish absolutely all claim and intention to use or acquire for burial purposes any other land in the Town of Lake View."[6] At the same

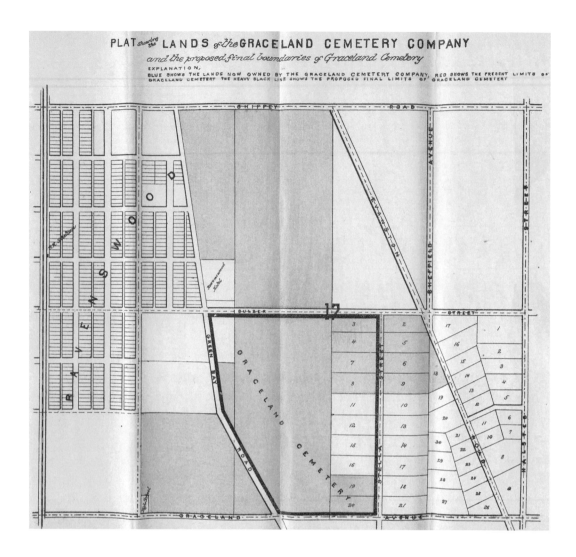

PLAT showing the LANDS of the GRACELAND CEMETERY COMPANY
and the proposed final boundaries of Graceland Cemetery
EXPLANATION,
BLUE SHOWS THE LANDS NOW OWNED BY THE GRACELAND CEMETERY COMPANY, RED SHOWS THE PRESENT LIMITS OF GRACELAND CEMETERY THE HEAVY BLACK LINE SHOWS THE PROPOSED FINAL LIMITS OF GRACELAND CEMETERY

6.1. Plat of Graceland's final boundaries, the outcome of the cemetery's compromise with adjoining Lake View (1879).

time, Graceland would sell its land located beyond the new limits. The cemetery also proposed to extend Stella Street (now Kenmore Avenue) north from Graceland Avenue (now Irving Park Road) and connect it with Sulzer Street (now Montrose Avenue). It would keep a fifty-foot margin along the western side of Stella Street "free from actual interments" and lined with "a hedge or row of trees."[7] In short, Lathrop offered to restrict the extension to 35 acres, rather than the original 190, and to forgo any future expansion. There was considerable financial incentive for Graceland to endorse the proposition, as they would be able to subdivide the extra land and sell it for residential development. The cemetery now acquiesced to the 1878 Supreme Court ruling and offered to "pay all taxes levied or assessed upon its lands, not now enclosed for Cemetery purposes, for the years 1875, 1876, 1877 and 1878."[8]

That March, some seventy Lake View voters petitioned the board of trustees that was the town's governing body to accept the compromise.[9] Edwin H. Sheldon, identified not as a Graceland officer but only as affiliated with Ogden, Sheldon & Co., was among the signatories. John A. Cole, a civil engineer, was another petitioner; Cole would gain employment at the cemetery only months later, perhaps at Sheldon's urging.[10] Even the two owners of property immediately adjoining Graceland, James Waller and Frederick Sulzer—"the only individuals [who] could be claimed to be specially affected"—supported the compromise.[11] Sulzer may have acted in part out of self-interest, however; his late father had earlier sold property to Graceland, and now, should the proposal be accepted, he hoped to do the same. After winning endorsement in a Lake View election, the compromise was officially approved on April 7, 1879.[12] With its boundaries now fixed in their present location, the cemetery's total area increased to 125 acres, as it remains today. In the coming years, the 150 acres of cemetery-owned land beyond the final limits, north of Sulzer Street (Montrose Avenue) and west of Green Bay Road (Clark Street), would be subdivided for residential development under the direction of Graceland's O. C. Simonds.[13]

THE SULZER TRACT AND JOHN A. COLE

Now the cemetery made its final land acquisition, purchasing a property from Frederick Sulzer.[14] Formerly known as the Sulzer Tract, it was located at the southeast corner of the intersection of Green Bay Road with Sulzer Street. Unlike Graceland's earlier acquisitions, this new area was already landscaped; it had been the homestead of Lake View's first white settler, Conrad Sulzer (1807–1873).[15] A native of Switzerland, Sulzer came from New York to Chicago in 1836. After a two-year reconnaissance, he purchased the land from developer William Ogden, at that time the city's first mayor and later one of Graceland's founding officers. Sulzer built a two-story farmhouse and cleared land to graze cattle. An avid horticulturist, he also began cultivating gardens and orchards. Although the extent of his original purchase is uncertain, after another acquisition in 1847 Sulzer's farm apparently embraced a hundred acres. By the time his son sold the homestead parcel to Graceland, much of the land had already been subdivided; some of it was now the suburb of Ravenswood. The cemetery itself had earlier persuaded Conrad Sulzer to subdivide, and in the first expansion of 1861 it acquired forty-five acres of his farmland. After Graceland's second purchase, Sulzer's farmhouse was moved to the southwest corner of the two roads, leaving behind the surrounding garden and other plantings.

Unlike the comparatively blank canvas on which Jenney and Simonds were at work, Sulzer's land was virtually an improved parkland. The map prepared by Charles Rascher around 1878 shows the general layout and plantings just after Sulzer's dwelling was removed. Adorned with deciduous and evergreen trees and shrubs, the improved land was apparently the handiwork of not only Conrad Sulzer but also his son Frederick, who was a florist, nurseryman, and landscape gardener.[16] Graceland now required a professional to redesign the property, to unite it both spatially and horticulturally with the broader cemetery. For reasons unknown, the board did not simply further expand the scope of Jenney's project, but commissioned Lake View resident John Cole for the job.

A New Hampshire native, John A. Cole (1838–1932) apparently gained a civil engineering education through apprenticeship.[17] After completing an academic course at Kimball Union Academy, a private school in Meriden, New Hampshire, around 1855 he entered the office of an unidentified "noted Consulting Engineer of Boston, then engaged in building the Hoosac [railroad] Tunnel." Three years later he was at work, possibly employed by the state of Massachusetts, on water supply projects at Sudbury Meadows and the Mystic Waterworks in Somerville. At the outbreak of the Civil War, Cole, a man of spiritual conviction, served on the U.S. Christian Commission, "a voluntary organization for carrying on religious work among the soldiers and aid to the wounded and dying."[18] After the war he established an engineering practice in Washington, D.C.[19] In 1872 he moved west and did the same.

Arriving in Chicago after the Great Fire, Cole won commissions to design Lake View's pumping station (1875) and, further south, the Hyde Park Waterworks (1881), including its mile-long intake tunnel beneath Lake Michigan.[20] He also designed the two towns' sewerage systems and contributed to Chicago's "Lake Shore protection works."[21] By around 1879 Cole had apparently achieved local renown for his skill with drainage and other hydraulic systems; Ann Durkin Keating notes that he was the official consulting engineer for Jefferson, Hyde Park, and Lake View by the 1880s, involved in the waterworks and other projects in these towns.[22] Cole's skill as a hydraulic engineer was likely not the main source of Graceland's interest in him, however; the Sulzer Tract, unlike the cemetery's low-lying eastern lands, was comparatively elevated and presumably more easily drained, requiring no advanced engineering. Frederick Sulzer, the tract's former owner, more likely links Cole to Graceland through connections forged in Lake View. Sulzer was then serving as Lake View's town clerk, and only months before the town had again employed Cole to modify its waterworks.[23] Both participated in the Graceland negotiations. Sulzer and Cole also had a mutual professional interest in landscape gardening. Four years earlier, for instance, the two had joined others to form the Ravenswood Improvement Association. The group's aim was to advance this nearby village, much of which was on Conrad Sulzer's former land, through "the improvement and beautifying of streets and parks" and "the planting and care of trees and foliage upon public and private property."[24] Frederick Sulzer, possibly together with Edwin Sheldon, may have influenced Bryan Lathrop's decision to retain Cole, perhaps as a political gesture of good will toward Lake View.

In his design revisions, Cole expanded Section R, laid out earlier by H. W. S. Cleveland, to fill the former Sulzer Tract. (Fig. 6.2) To access the new burial lands, the engineer projected a thoroughfare extension off of the cemetery's Western Avenue. Cole's new road, called Cedar Avenue, followed a curvilinear trajectory and terminated in a roundabout, its plan resembling an inverted water droplet. In the center of the roundabout Cole designed a circular lot array. He also configured a crescent-shaped band of plots to follow the inside of the new drive's curve. Otherwise, save for the plots lining the new north and west limits of the expanded section, he organized the majority of the plots and attendant walks in Section R with a grid aligned perpendicular to Green Bay Road. In comparison to the fluid geometries of the layout Jenney would soon produce for the cemetery's eastern lands, Cole's extension is more rigid. And whether Cole's design abilities included horticulture is uncertain. Apparently at least some, if not most, of Conrad and

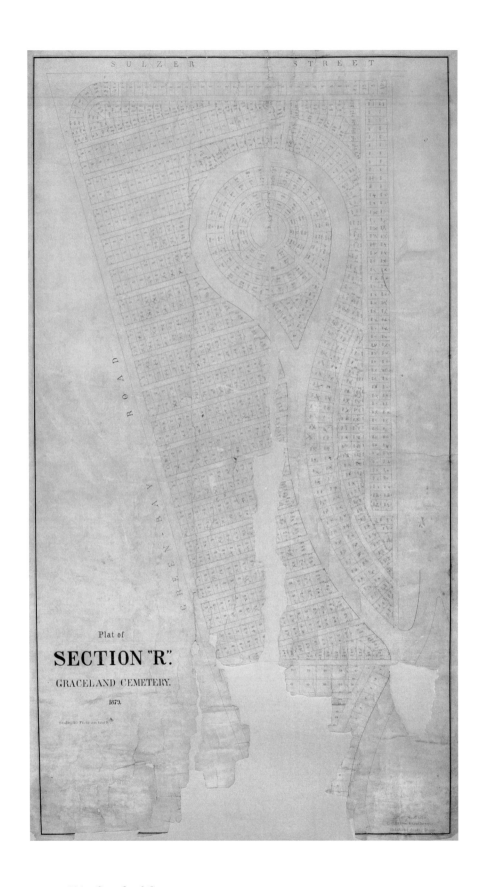

Frederick Sulzer's earlier plantings, certainly mature by 1879, were retained.[25] Despite Jenney's presence at Graceland, Cole apparently undertook the work independently; no evidence of connection or collaboration between the two has come to light. Moreover, Cole's work was not confined to the Sulzer Tract. Some weeks in advance of the Lake View compromise, he was at work in Graceland's Section C, completing cross-sectional surveys of it that March.[26]

(*Opposite*) **6.2. John A. Cole's layout of Section R (1879).** Courtesy Chicago History Museum (ICHi-52321).

GRACELAND EXPANDS EAST

Along with the Sulzer Tract acquisition, the Lake View agreement permitted Graceland to convert unimproved land within its new limits to burial use. The area now to be developed lay within a section of what was formerly known as "Iglehart's Sub-division."[27] The proposition document describes it as a block of contiguous lots numbered, from north to south, 3, 4, 7, 8, 11, 12, 15, and 16. Graceland already owned lots 3–8, as well as 19 and 20, adjoining at the south (the last two had been improved earlier).[28] After the success of the compromise, the cemetery now purchased Iglehart lots 11–16. The map accompanying the printed proposition shows that the resulting rectangle of land abutted the entire length of the cemetery's eastern boundary, the soon-to-be-extended Stella Street. Jenney's canvas was set to expand.

Early in 1879, the board of managers instructed Jenney to prepare a layout for not only "the low lands" but now also "the eastern portion of the grounds embraced by the new limits of Graceland." "Low lands" refers to the area Jenney had already begun draining the previous autumn, and "eastern portion of the grounds" to the former Iglehart properties. The board accepted his completed plan for both areas in early May.[29] As the cemetery's final boundaries had been fixed only weeks earlier with the successful Lake View compromise in April, it might appear that the landscape gardener responded very swiftly indeed. It is more likely, however, that Bryan Lathrop, anticipating a favorable resolution with Lake View, whether by litigation or negotiation, had instructed Jenney to consider this portion of the cemetery at the outset of the project in 1877.

Like those of his predecessors, Jenney's Graceland plans and other drawings have been lost. Charles Rascher's project to map the cemetery, although begun around the time Jenney was com-

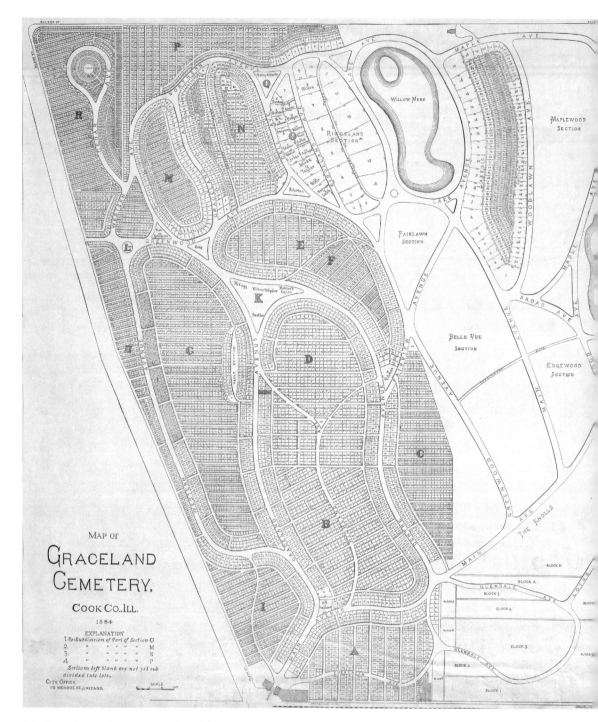

6.3. Graceland's layout and subdivisions in 1884.

Courtesy Chicago History Museum (ICHi-31174).

missioned, was apparently finished in advance of the landscape gardener's new layout, missing the opportunity to record it perhaps only by months. After Rascher's map, the next known graphic representation of the cemetery's layout is an 1884 plat titled *Map of Graceland Cemetery, Cook Co., Ill.*, and it is the main source of our knowledge of Jenney's design.[30] (Fig. 6.3) The fundamental purpose of this map was to document the cemetery's drives, walks, section subdivisions, building locations, and water features as they existed at that time. It records neither vegetation nor topography.

The names recorded on the map are as important as its graphic representations, because they reflect a decision, presumably the board's, to employ a new nomenclature scheme for the cemetery's component parts, one that effectively distinguishes Jenney's work from that of his predecessors. The map shows that Graceland ceased identifying its burial sections with alphabetic labels and adopted names evocative of each new section's locale or setting, such as Ridgeland, Edgewood, and The Knolls. Roads and walks, and even some of Graceland's older drives, also received new names. A northern segment of Lake Avenue is now given a more directionally accurate name, Northern Avenue. Glenwood has become Greenwood, although it is unclear whether this was meant to evoke Graceland's New York counterpart or the character of the surrounding vegetation, and Woodlawn is similarly ambiguous. Eastern Avenue, made a misnomer by the cemetery's expansion, has now become Centre (later spelled Center) Avenue. These more descriptive place names reflect the broader landscape ethos that pervaded Lathrop's, if not the entire board's, thinking at that time. They also allow for a more nuanced appreciation of Graceland's landscape, whether as it actually was or as projected in Jenney's design. Even though the map did not show topography or vegetation, names like Edgewood and Ridgeland enable us to conjure up images of the landscape setting of these sections.

It is unclear who was responsible for devising these new names, but it was likely Jenney. He had already conceived a similar typology to identify the drives in his layout for Riverside Cemetery in Moline in 1874. Two of Graceland's names in particular suggest that it was Jenney who coined them. The first is Lotus Avenue. Not unlike his former employee John Edelman's fascination with "Lotus of the Calumet" (see chapter 7), Jenney apparently found a similar allure in the indigenous plant. The lotus was also a popular motif for cemetery monuments; its association with the afterlife derived from its use in ancient Egyptian design, a style source thought to be appropriate for funerary architecture.[31] One is tempted to imagine that this aquatic species once adorned the surface of nearby Lotus Pond, complementing the lotuses that ornament many of Graceland's monuments. The second, Belle Vue (subsequently spelled Bellevue), is a name that bespeaks the hand of a Francophile like Jenney.

As intriguing as this map is, it is of limited use in assessing Jenney's design contributions at the cemetery. In 1880 Jenney's assistant, O. C. Simonds, would supplant him as Graceland's landscape gardener. The map was published four years later; thus it is difficult to distinguish Jenney's changes from those that might have been made later by his successor or by Bryan Lathrop. But the bones of Jenney's layout can be recovered by considering the 1884 plat alongside Rascher's map and complementing these with textual sources and stylistic evidence.

Jenney transformed almost the entire lowland area into an artificial lake. After completing the sewer in the spring of 1879, he excavated the newly drained slough to make the lake, later named Willowmere.[32] Its creation was the first phase in the implementation of his overall layout. The cemetery's desire for such a lake was a longstanding one; as we saw in chapter 4, H. W. S. Cleveland had apparently projected a similar feature nearly a decade earlier. Now at last constructed under Simonds's supervision, the new lake was not intended to be merely decorative. Jenney's design, like its earlier West Parks counterparts, fused beauty with utility, for the purpose of Willowmere was not only to ornament the grounds, but to drain them, functioning as a reservoir.[33] (Fig. 6.4) Fulfilling Bryan Lathrop's aesthetic expectations—undoubtedly of the picturesque—Jenney configured the lake with a sinuous outline and embellished it with an island retreat. (Fig.

(*Below and opposite*)
6.4. Jenney's plan of Lake Willowmere and outlying burial sections (c. 1878). Although unsigned, this drawing is possibly in Jenney's own hand or that of his employee O. C. Simonds.
Courtesy Chicago History Museum (ICHi-61227).

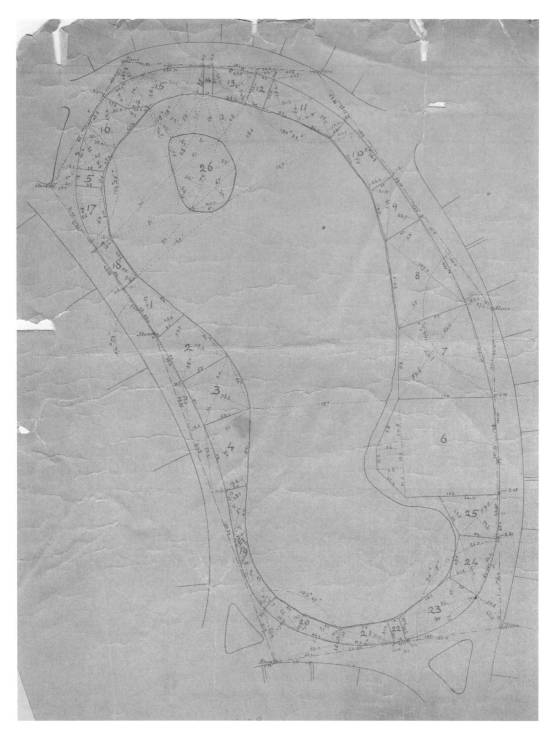

6.5. Survey of Lake Willowmere and its island (c. 1878).
Courtesy Chicago History Museum (ICHi-61225).

6.5) A period photograph reveals that the island included a wooden gazebo; Jenney had designed a stylistically similar structure for the Riverside Hotel.[34] (Fig. 6.6) Next, using the sand and gravel excavated to create the lake, Jenney elevated its surrounds and molded the terrain in accordance with his new design. Although we no longer have his topographical contour plans, texts confirm that the earthworks Jenney undertook were substantial. Simonds, for instance, identified one of the new landforms as a hill "ten feet high with a steep slope in one direction and a long tapering slope in another," which was given a top layer of good soil and planted.[35] Lake Willowmere and its island would soon become a visual fulcrum within Graceland's evolving landscape composition, and today they remain one of its most celebrated features.

Jenney's plan extended the cemetery eastward into the undeveloped Iglehart lots. Here the land was also poorly drained and swampy, so much so that excavation and earthworks were soon to be given top priority. Planting would wait. As he had similarly done first at the "low lands," Jenney proposed to remedy the problematic drainage with two artificial water bodies, now long vanished.

6.6. View across Lake Willowmere to the island around 1880.
Courtesy Chicago History Museum (ICHi-22705).

As represented in the 1884 plat, Hazel Mere (later spelled Hazelmere) and Lotus Pond were diminutive in scale relative to Lake Willowmere. Both were located near the cemetery's eastern Stella Street boundary; Hazel Mere was located approximately in lot 8 of the Iglehart tract and Lotus Pond in lot 16, closer to Graceland's southeast corner. A later source alludes to the pair and suggests that they were constructed around the same time, shortly after Willowmere, but apart from that we know nothing about how these lakes were created, or by whom.[36] But the stylistic compatibility of their extruded oblong forms with the outline of Willowmere suggests Jenney's hand.

Along with Hazel Mere, Lotus Pond, and the associated earthworks, textual evidence tells us that Jenney designed their wider frame. The board's meeting minutes most reliably confirm that roads were among the proposed improvements depicted in Jenney's plat. Although this document no longer exists, his layout can be discerned by juxtaposing the 1884 Graceland plat with the map Charles Rascher made around 1878. Comparing the two reveals that Jenney planned to convert the soon-to-be-drained area into six major burial sections, organizing the precinct with a circuitous drive. In turn, this thoroughfare delineates the oblong bounds of the new sections. In his scheme, three sections now filled the cemetery's northeast corner: Lakeside, located directly east of Willowmere, is bounded by Maple, Woodlawn, Main, and Lake avenues; adjoining it is Maplewood, defined by Maple and Woodlawn avenues; and the third section, later named Thorndale, is delineated by Broad, Wildwood, and Maple avenues. The central portion of the new eastern expansion also gained a trio of sections: Fairlawn, Belle Vue, and Edgewood, each bounded by three avenues. The Knolls, a new section bounded by Main, Glendale, and Lotus avenues, now occupied Graceland's southeast corner.

Overlaying the 1884 plat atop Rascher's map indicates that Jenney's plan was not confined to Graceland's new eastern half. At the interface of his layout with the existing cemetery, he reshaped Cleveland's sections and roads, joining them together to form one cohesive composition. Erasing their original outlines, Jenney now subsumed Section T within his new Belle Vue area and Section U within Fairlawn. He also expanded Section C, altering and shifting its boundary further east. This section, unlike the others Jenney modified, retained its original name in the new layout. He also reconfigured three of his predecessor's drive intersections in order to integrate them seamlessly into his new road network. The first was Section W and its adjoining roundabout, located at the junction of Lake and Northern avenues, near Lake Willowmere. Here he refashioned them to make a new triangular or wedge-shaped three-way intersection, allocating the land within the triangle for burials. This diminutive section would later be called Getty, after its purchaser. Jenney similarly modified the intersection of Dell Avenue and Greenwood. There he sliced a new path across the southeast tip of Section N, creating another triangular plot, labeled on the map with its owner's name, Allerton. He treated the junction below Section V (renamed Ridgeland) similarly; the new plot itself was later named Kranz, also after its owner. Jenney also applied this method of creating a triangular burial space at the intersection of three roads within the new cemetery extension itself. He created one section, later identified as Bosch, at the convergence of Lake, Main, and Fairview avenues, and another bounded by Broad, Wildwood, and Maple. Located west of Hazel Mere, this latter section was the largest of the triangular type and later became known as Thorndale. Of his entire constellation of interstitial plots, the metamorphosed Section W would later become the most distinguished, with a jewel-like monument—the Getty tomb—designed by Jenney's former employee, Louis Sullivan.

JENNEY'S DESIGN SOURCES

When Jenney decided to employ artificial lakes as drainage reservoirs at Graceland, he was reapplying a technique he had already long perfected. The configurations of the cemetery's lakes, however, are a departure from the markedly irregular, twisted shorelines of Jenney's earlier West Parks lagoons. Willowmere's margins, for instance, follow a bolder, more simplified curvature. When seeking design inspiration, Jenney did not turn directly to English sources as Lathrop would likely have done. For Graceland, it seems, the landscape gardener instead adapted a form vocabulary that drew on picturesque examples indirectly, through a French filter.[37]

The English gardener and critic William Robinson, as Jenney's biographer Theodore Turak observes, distinguished a number of attributes unique to French landscape gardening in his study *The Parks, Promenades, & Gardens of Paris*. First published in 1869,

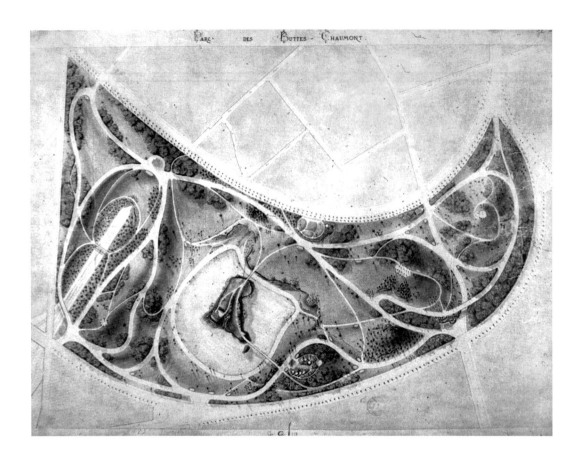

6.7. Plan of Parc des Buttes-Chaumont, Paris, designed by Jean C. A. Alphand (1865–1867).
Hulton Archive/ Getty Images. Used by permission.

Robinson's monolithic survey—more than six hundred pages in length—was probably known both to Jenney and to Bryan Lathrop. French designers, according to Robinson, first composed drives and paths in "tight geometric configurations which frequently took the form of ellipses and sections of circles" and then topographically adjusted them.[38] Measuring French parks against their English counterparts, Robinson lauded the former's "boldness and finish." Indeed, the Englishman's commentary, as Turak discerned, gives one the impression that he believed the "French had surpassed the English in their own garden type." Jenney's Lake Willowmere configuration is stylistically resonant with the fluid, elegant curvilinear geometries of such parks as the French landscape gardener Jean C. A. Alphand's Parc des Buttes-Chaumont (1865–1867). (Fig. 6.7)

Alphand, Turak writes, "designed an informal garden with the same studied care Le Nôtre used in laying out the grounds of Versailles in geometric patterns," and Jenney would do no less at Graceland.[39] Although Buttes-Chaumont was constructed after

Jenney's time in Paris, the Francophile landscape gardener would likely have known it from literary sources. Robinson, for instance, devoted a chapter to the park and believed it to be the "most interesting garden in Paris."[40] And, as we will see, Willowmere's island would later become a renowned burial site, although it is unlikely that Jenney intended it for such a use. But with the addition of a boulder monument his lake-and-island ensemble became evocative of another French precedent, Ermenonville. Laid out in 1762 by the marquis de Girardin, this iconic picturesque garden similarly included a lake with an island, the Île aux Peupliers (Isle of Poplars), among its features. In 1778, Ermenonville gained a new patina of meaning when the celebrated philosopher Jean-Jacques Rousseau was interred on the island.[41] (Fig. 6.8)

As we've seen, when Jenney was designing Riverside Cemetery in Moline, the comparatively rugged terrain, together with a concern for economy of construction, compelled a curvilinear layout. Graceland's low lands, in marked contrast, now offered him a tabula rasa. Yet despite the flat topography, Jenney again elected to compose the grounds using a sinuous form template. What were his sources? As we are about to discover, Jenney's "city of the dead" drew much from designs for towns to be occupied by the living. Most immediately, his layout is indebted to his intimate knowledge of the spatial devices Olmsted and Vaux employed in

6.8. Ermenonville, Jean-Jacques Rousseau's cenotaph by Hubert Robert on the Isle of Poplars.
Cenotaph of Jean-Jacques Rousseau, Ermenonville. Photograph by Parisette, 2006. Wikimedia Commons.

designing the town of Riverside. Although the two layouts are by no means mirror images, there is a resemblance, and it is not accidental; Jenney seems to have adapted some of Olmsted and Vaux's Riverside design aspirations and transferred them to the cemetery. At Riverside, as Walter Creese writes, the "impression of dwelling beneath a low green cloud of foliage to protect the premises from the brightness of the prairie sky was carefully cultivated," and "the slow, rhythmic turns" of the suburb's winding roads meant that its "inner space of 1,600 acres might appear to be almost infinitely varied." He concludes that "the ambience achieved the maximum effect with the minimal means of foliage and curving roads, but with extraordinary domestic architecture all along the way."[42] There is little doubt that Jenney hoped to realize analogous landscape effects at Graceland. His cemetery drives, for instance, turned no less rhythmically, the landscape orchestrated to unfold as the viewer traveled through it. In contrast to today's automobile experience of Graceland, moving at the pace of a horse-drawn carriage, or even on foot, must have intensified the spatial sensation. Instead of Riverside's "extraordinary domestic architecture," Graceland would offer monuments, some no less remarkable, as surrogates for villas and cottages. Another key aspect of the suburb's design, as Creese also observes, was that "Olmsted laid a carpet of continuous green, accepting the prairie flatness, and even sinking the streets two [to] three feet in order to keep the basic earth plane moving."[43] Jenney too valued the import of the terrain, and the new landforms he engineered at Graceland would also gain a lawn mantle and similarly pulsate and swell within their carriage-drive frames.

Olmsted and Vaux's Riverside, however, was not Jenney's only design source. He also apparently considered another, even earlier suburban precedent—this one on the outskirts of Paris. Le Vésinet, situated on a rail line like its Chicago counterpart, is located about seven miles northwest of Paris on what was previously a royal hunting preserve. (Fig. 6.9) In 1855 the landscape architect Paul de Lavenne, the comte de Choulot, began to transform the forested site into a town, recasting it in the picturesque mode. Turak writes:

> Choulot described how he penetrated into the forest, or as he said, into the picture. He noted the rise and fall of land and

tried to isolate significant points of the landscape. The eye would be conducted to these points—hills, lakes, or buildings. The choices offered by the landscape were infinite, but the final determinant was light. He suggested that perspectives be placed diagonally to the path of the sun whenever possible. In this way the configurations of landscape appeared rich and plastic.

Even today, walking through Le Vésinet is a marvellous experience. One's eye is guided through wide, long spaces that the count called *coulées*. Simply translated, *coulée* means path, but to Choulet, they were the essential elements of unity. . . . These swaths of lawn, positioned so carefully, were placed off-limits to all vehicular traffic. They were intended for walking, cycling, and playing.[44]

As the coulées suggest, there are many affinities between Le Vésinet and Riverside. Indeed, Turak compellingly argues that the French layout "provided much of Riverside's inspiration," contending that these "great sweeping areas of grass" are the "most striking similarity

6.9. View of Le Vésinet. William LeBaron Jenney would draw on the layout of this Parisian suburb as a design source. Courtesy www.bellecpa.com.

between the two."[45] Le Vésinet's coulées and Riverside's "Long Common," for Turak, "differ only in nuance"; the coulées "are broader with an almost baroque feeling," while the Riverside parkland's "more and closely spaced trees tend toward a sense of intimacy." As Turak concedes, however, there are significant differences: "Like the French parks," Le Vésinet was "more tightly planned" than its American counterpart. In contrast to the "amoeba-like" blocks delineated by Riverside's curving streets, in Le Vésinet the drives follow "more disciplined curves." Curiously, this could also be said to distinguish Graceland from Riverside; the cemetery's drives are arguably more disciplined than those of the suburb.

While he was still working at Graceland, Jenney's landscape gardening talents attracted French recognition. In 1879, the celebrated landscape gardener Edouard André (1840–1911) endorsed Jenney's work at Riverside in his treatise *L'art des jardins: traité général de la composition des parcs et jardins.* Tellingly, perhaps because Jenney's contributions there were so extensive, André mistakenly attributed the layout of Riverside only to "professeur Jenney," not Olmsted and Vaux.[46] Moreover, Riverside was the only Chicago example included in André's listing of thirty exemplary American private parks and gardens. Coincidentally, elsewhere in the volume he does not fail to praise Adolph Strauch's achievements at Spring Grove.[47]

At least twice in later years, Jenney, presented with a substantial landscape gardening commission, again drew on the French-inspired form vocabulary he apparently first developed while designing Graceland. In 1882 the Lake Forest Cemetery Commission engaged him to redesign that city's cemetery.[48] (Fig. 6.10) Perhaps this job came his way because then, unlike today, Jenney was known as Graceland's author. Positioned on the lake some thirty miles north of Chicago, Lake Forest occupies gently rolling, ravine-traversed, and heavily wooded terrain. In 1857 the landscape engineer Almerin Hotchkiss composed the projected town (which was incorporated as a city in 1861) in a "romantic picturesque street plan."[49] He also identified a bluff-top, waterside location for its cemetery, a twenty-three-acre site affording picturesque views not only of the lake below, but also inland across the undulating topography. Remarkably, Hotchkiss himself was already well versed in cemetery design, perhaps even more so than in the layout of new towns; he had earlier made design contributions to the expansion of Green-

PLAT
--- of a portion of ---
LAKE FOREST CEMETERY,
otherwise known as
— Evergreen Cemetery —
Being a part of Lots One Two Three and Seven of the
Original Village of Lake Forest.

Adopted by the City Council of the City of
Lake Forest April 3rd 1882

Sylvester Lind
Attest   Mayor of the City of Lake Forest
George Fraser
City Clerk

Scale Sixty feet to one inch

UNIVERSITY AVENUE.

LAKE AVENUE.

6.10. Jenney's layout of Lake Forest Cemetery, north of Chicago (1882).
Courtesy Special Collections, Donnelley and Lee Library, Lake Forest College.

Wood Cemetery in Brooklyn, New York, and had also laid out two new rural cemeteries, Bellefontaine in his own St. Louis and, back in Illinois, Rock Island's Chippiannock Cemetery. The latter, located not far from his Riverside Cemetery in Moline, was probably known to Jenney. The site he selected at Lake Forest was not put to burial use until 1860, when the Forest Cemetery Association retained a local civil engineer who laid the site out in gardenesque-inspired "irregular, unrelated forms."[50] In 1881 the cemetery was acquired by the city and renamed Lake Forest Cemetery. Taking one of its first actions the next year, the city commission that ran the cemetery hired Jenney to redesign it. Discounting the original plat, Jenney responded instead with a new, more fluid layout. In his scheme, the cemetery's confining perimeter drive follows a curvilinear path apparently dictated by the topography. The inner paths adhere to more elegant curves and interlink the layout's circular centerpiece. Rather than following contours, these are evocative of the Francophilic lines of Graceland's drives. Along with a new drive and path network, he also relocated the cemetery's entrance and proposed a new chapel. Apparently planning to implement "phased development," Jenney platted only a portion of the cemetery for burial use, allowing the city commission to meet Lake Forest's changing needs through time.[51] As he already knew, such a development strategy was not unlike the one Thomas Bryan had employed at Graceland.

A second opportunity for Jenney to sculpt French-inspired landscape forms came six years later. In 1888 he was commissioned to design a Winter Garden conservatory in Chicago's Douglas Park (1871). (Fig. 6.11) Jenney had designed the wider parkland setting of the projected building almost twenty years earlier. Unlike the other West Parks, Douglas Park was decidedly focused on water—the lake Jenney had created when he laid out the park—and he had intended to provide, as Reuben Rainey puts

6.11. Plan of Jenney's design for a Winter Garden conservatory and its surrounds in Douglas Park, Chicago (1888).
Courtesy Chicago History Museum (ICHi-63845).

it, "the opportunity to become more absorbed in the lush land-scape through a series of lakeside promenades with a changing series of water views."[52] Even in this earlier work, Rainey detects resonances with Alphand's park design technique.[53] Revisiting his earlier layout, Jenney now positioned the new conservatory lagoon-side. When laying out the building's surrounds, he took the opportunity to revise a segment of the lake's original shore-line, unknotting and simplifying its margin.[54] This act, if only metaphorically, extended Willowmere's influence, enlarging its historical significance. As it did earlier at Graceland, the land-scape gardener's French-inspired aesthetic had found a compat-ible home amidst, appropriately enough, a regional landscape type with a name of French origins, the prairies.

SEVEN

The Era of Bryan Lathrop
and O. C. Simonds

William Le Baron Jenney's decision to involve O. C. Simonds with the Graceland project would have profound and unforeseen consequences for both men. (Fig. 7.1) Indeed, apart from Graceland, it is unclear whether Jenney involved him with any other work in the office. In January 1880, nearly two years into the project, the cemetery's development had apparently reached a new crossroads. That month, Graceland resolved "to do the work described in the estimates of the Company's engineer," adding that it hoped the job would completed that year.[1] Although the document detailing the precise nature of this project is now lost, it was presumably the implementation of Jenney's layout of the cemetery's unimproved eastern lands. The expense of the work, nearly $200,000 in today's dollars, confirms it as a major undertaking. At the same time, the board also resolved to "clay and gravel" a new road and use its "discretion as to the number and cost of large trees to be planted in Graceland."[2] The latter qualification is significant for a number of reasons. First and foremost, it documents that Jenney's improvements included plantings, calling not just for trees, but for sizable ones. Second, the board's choice to reserve its discretion with respect to planting was probably the outcome of Bryan Lathrop's influence; Graceland's vice president was now poised to expand his own design input at the cemetery.

According to an unpublished memoir by Edward Renwick, who began working with Simonds in 1882, as the implementation of Jenney's layout progressed Lathrop increasingly found that "all questions had to be referred to Mr. Simonds," and this circum-

7.1. Ossian Cole Simonds (1855–1931). Courtesy Morton Arboretum, Lisle, Ill.

stance would ultimately lead him to dispense with Jenney's services. In the interim, however, the rapport between Lathrop and Jenney's assistant escalated. Presumably before ending Jenney's involvement, Lathrop apparently offered Simonds a job as Graceland's landscape gardener.[3] Accepting Lathrop's inducement and doubtless emboldened by it, he resigned from his former instructor's office in 1880. Having successfully used Jenney's office as a career springboard, Simonds now left a "father figure" to join Bryan Lathrop as a landscape gardener brother-in-arms in the campaign to transform Graceland's eastern half into a work of landscape art.[4] He was not the only one to depart; perhaps simultaneously and at Simonds's persuasion, William Holabird (1854–1923), another civil engineer with architectural ambition, also exited that year.[5] The pair now formed their own partnership, Holabird & Simonds.[6] As Lathrop apparently employed Simonds independently at the cemetery, he must have consented to his landscape gardener's parallel involvement in this new enterprise. Indeed, Simonds's Graceland post may well have provided him with the initial capital to establish the firm. Although the precise sequence of these episodes remains somewhat ambiguous, they were most certainly interconnected.

Jenney had now lost not only a substantial commission but also two of his staff. How did he respond to this presumably unforeseen outcome? One later account suggests that he withdrew amicably, effectively turning the project over to launch his former student's career.[7] Although somewhat doubtful, such a reaction does align with Jenney's "father figure" reputation. In a similar vein, there is nothing to suggest that Simonds's first job as the elder architect's assistant was meant to be a permanent one. On the other hand, the resignations more likely would have injured Jenney's morale, if not his practice. Of the two, William Holabird's probably inflicted the greater wound. Jenney had by now invested five years in training him; unlike Simonds, Holabird had abandoned university studies before completing them.[8] Especially if the pair quit at the same time, Jenney's reaction was amicable in only a cool professional sense. One also wonders what Thomas Bryan, apparently returned to Chicago from Washington, D.C., by 1880, would have made of his nephew's decision.

O. C. SIMONDS

In October 1881, Bryan Lathrop was elected Graceland's president. That year, in one of his first official acts in this capacity, he named Simonds the cemetery's "Superintendent, Landscape Gardener, Engineer, and Surveyor."[9] What motivated him to entrust Graceland's improvement to a twenty-six-year-old with only three years' experience, all of it confined to one project? In fact, Simonds's status as a relative novice was probably central to his appeal; the new president, whom Simonds later called a "most able amateur in landscape designing" himself, apparently made the choice predicated on his own supervisory role continuing.[10] Now free of Jenney's mediating influence and no longer content to passively oversee the work, Lathrop planned to deepen his involvement. By selecting a comparatively inexperienced designer who would need guidance, Lathrop opened the door for himself to assume Jenney's role. Indeed, Simonds later credited not Jenney but Lathrop as informing his decision to change his "life's work from architecture to landscaping."[11] Unlike his former employer or other experienced landscape gardeners, Simonds had yet to develop his own design ideals. Consequently, at this juncture, the plan now to be developed would be shaped more by Lathrop than Simonds. Indeed, the future landscape gardener later attributed much of the cemetery's beauty to Lathrop's "good taste and continued interest."[12]

Another factor is that the scope of Lathrop's anticipated landscape improvements was apparently expansive enough to require a designer prepared to devote years, not months, to Graceland. After more than a decade's experience with the cemetery, Lathrop knew all too well that a landscape gardener, in his words, "must plant with the eye of a prophet, for it requires many years to bring to perfection the picture he has imagined."[13] Fortuitously, Lathrop encountered Simonds at the very start of his career, and the enthusiastic novice probably appealed to him as offering a conduit to vicariously realize his own landscape gardening vision. Now all that remained for Lathrop to accomplish was to convince Simonds to forgo his architectural aspirations, and at this he would ultimately succeed. Throughout the coming years, as Simonds gained experience in his newly adopted pursuit, the development of Graceland would become an increasingly collaborative endeavor. Eventually, the renown gained by Simonds's own design contributions would eclipse that of all his predecessors, compelling a closer examination of his formative career experiences and his exit from architecture.

Past scholarship on Simonds has considered him almost exclusively through a landscape gardening lens.[14] When he entered the University of Michigan in 1874, however, Simonds chose to pursue civil engineering. That year, the university's catalog may well have first brought the possibility of an architectural career into his ken. "No technical course in Architecture has been established," the catalog advised, "but the course in Civil Engineering includes all the Principles of Drawing which are necessary for representing Designs and working out Detailed Drawings; and the Lectures on the Resistance of Materials and Stresses on Frames are sufficiently comprehensive to enable the student properly to proportion a Structure. A special course of reading of works in the Library is also marked out for those who desire to pursue this study."[15] The absence of an architecture program reflects the national dearth of professional courses in the subject at that time. This circumstance soon changed, at least at Michigan; in 1876, two years into Simonds's studies, the university hired William Le Baron Jenney to run its newly established Department of Architecture and Design.[16]

Simonds promptly transferred into the new curriculum and now "expected to go into architecture as a profession."[17]

According to his Michigan transcript, Simonds began architecture in his junior year, 1876–77.[18] That first semester his courses were Free-hand Drawing and History of Architecture, along with physics and calculus. He continued in the next with the second part of the history survey, Geometrical Drawing, Analytical Mechanics, and Elementary Architecture, along with chemistry. Jenney's course descriptions, as recorded in the university's register for 1876–77, further reveal the nature and scope of Simonds's formal architectural education. The course titled Free-hand Drawing, for instance, embraced not only "sketching from nature," but also the "application of natural forms to ornamentation." Geometrical Drawing complemented free-hand technique with "plane projection" in ink and watercolor and "ornamentation and lettering." The history of architecture course include an overview of examples "from the earliest period to the present date, illustrated by sketches, photographs, models, etc."; its two-semester duration suggests that Jenney valued historical knowledge as foundational to contemporary practice. More pragmatically, Elementary Architecture included "carpentry" and "stairs" among its topics.[19]

Although he later claimed that he knew nothing about landscape gardening until after he came to Chicago, Simonds probably first gained awareness of this alternate profession while at Michigan. Certainly we can expect that Jenney took his lectures as an opportunity to expose students to his own works not only in architecture but also in landscape gardening, such as his layout of Chicago's West Park system. More directly, Simonds's surveying class, Analytical Mechanics, provided instruction in "setting out topographical work" such as "walks, drives, [and] lakes."[20] Most explicitly, Jenney in fact offered a course titled Landscape Architecture, described in the *General Register* as featuring "designs for private and public grounds, and the architectural details."[21] Simonds did not choose to enroll in it, which likely indicates his resolve to become an architect. Unfortunately for the aspiring architect and his instructor, the new department was terminated the next year.[22] No doubt disappointed, Simonds returned to finish the engineering degree his senior year. Like Jenney himself and possibly even at his suggestion, he set out to make civil engineering his gateway into architecture.

After Simonds graduated in June 1878, he gained early professional experience surveying Lake Michigan's east shore for the U.S. Geodetic Survey.[23] Ready by autumn to begin his career as an architect, he approached his former instructor. Beyond the classroom, Jenney was known for helping his students enter the profession by giving them their first jobs. He did no less for Simonds, employing him as an assistant.[24] As a fellow Michigan alumnus, the architect Irving K. Pond (1857–1937), put it, the new graduate would now "use Major Jenney's office as a place from which to 'take off.'"[25] Pond soon did the same.

Arriving in Chicago in the autumn of 1878, Simonds was soon lodging with a Michigan classmate, Clarence O. Arey (1857–1896).[26] Simonds's arrival was timely; Chicago was then on the cusp of an economic resurgence, as the long depression that followed the Panic of 1873 was at last drawing to a close. Irving Pond joined Arey and Simonds the following August; he too was to begin working for Jenney. He later recorded his recollections of the architect's practice: "Just at the time [Simonds and Arey] came to Chicago Major Jenney was acting as landscapist for Graceland Cemetery as well as for others of his clients."[27] This is a valuable piece of evidence, for it indicates that Jenney had secured the Graceland commission before hiring Simonds. Indeed, perhaps it was this new work that now enabled him to employ his former student. Although Simonds later claimed that Jenney's cemetery input was limited to engineering activities, Pond's account suggests otherwise.[28] Calling Jenney a "landscapist" suggests that he was doing far more at Graceland than excavations, drainage, and earthworks. No less illuminating is Pond's recollection that Jenney had clients—presumably landscape clients—other than Graceland. Jenney's highly visible work at Chicago's West Parks and at Riverside had likely attracted the new jobs. At the time Simonds arrived at the architect's office, for Pond at least, the amount of landscape work then on the drafting boards was memorable. This makes it all the more perplexing that Simonds later claimed ignorance of landscape gardening until Bryan Lathrop introduced him to it.[29] Nonetheless, it is difficult to believe that Jenney was silent on the subject, either now or earlier in the classroom, with the new employee to whom he was about to assign a major undertaking in landscape gardening.

Pond writes that after Simonds had been working in the office for a few weeks, Jenney dispatched him to Graceland "to look after certain matters which needed immediate attention."[30] The project, Simonds later recounted, "called for some engineering skill in putting in drains, excavating a lake, grading and building roads, and in grading the various sections."[31] His new assistant must have early demonstrated enough promise to win Jenney's confidence that he was up to the task. Many years later, Simonds self-effacingly explained that "knowledge gained in acquiring the degree of civil engineer . . . made me somewhat useful to the cemetery company" and to Jenney.[32] Once at Graceland, Simonds soon, in Pond's words, "made good and was appointed Major Jenney's representative on the work."[33] Even so, according to Simonds himself, he was disappointed with the post at first.[34] He had, after all, come to Chicago to pursue a career in architecture. Now the frustrated architect found himself "shunted," as Pond termed it, into landscape gardening.[35]

Simonds later named his pastoral childhood in rural Grand Rapids as one of the considerations that informed his choice to redirect his career from architecture to landscape gardening. Throughout his "early life on a farm," for instance, he "studied botany and made considerable acquaintance with the native growth."[36] And yet then, as now, esteem for the natural world and the plants that grow in it was not confined to landscape gardeners. It could also be found within Chicago's architectural scene, especially at Jenney's office. Far from ordinary, it was then gaining recognition as an incubator of innovation, culminating with the design of the Home Insurance Building, the first steel-framed skyscraper (1885). And yet Jenney's innovative outlook was not purely technological in nature.[37] For him, the relevance of the natural world to design was not limited to landscape gardening; it also informed his architecture. Most directly, he saw it as a form source for ornament.[38] Indeed, he had already inculcated his new assistant with his botanically derived ornamentalist technique earlier, when Simonds took his Free-hand Drawing class at Michigan.

Louis H. Sullivan was another aspiring architect who began his Chicago career with Jenney. Gravitating to his office in 1873, only five years before Simonds, Sullivan soon befriended the architect's foreman, John Edelmann.[39] Both architects held the natural world as a profoundly inspirational font, and Edelmann became the new arrival's first intellectual mentor. Less than a year later, however, Sullivan left to undertake study in Paris. When he returned to Chicago in 1875, he began practicing as a freelance designer. Sullivan's professional whereabouts during Simonds's tenure with Jenney from 1878 to 1880 are unclear, but apparently he held a series of short-term jobs.[40] Nonetheless, were he in need of work or perhaps only to make a social call, Sullivan may have visited his former employer and possibly met the new staff member. Edelmann's presence in Jenney's office might also have lured Sullivan back. According to Pond, Clarence Arey, too, "discovered" Edelmann there, and soon Sullivan's muse became "a pretty constant visitor" to the apartment the two shared with Simonds "With Simonds," Pond revealed, Edelmann "talked transcendentalism of a sort."[41] That a Transcendentalist ethos pervaded Edelmann and Sullivan's architectural thought is well known.[42] But Pond's reminiscence is revelatory; it is perhaps the only source to document Simonds's own awareness of, and presumably interest in, the subject. Pond chronicled another exchange between Edelmann and the trio of aspiring architects:

> While John was talking all his happy nonsense to me and my companions, he would be using his pencil semi-consciously on scraps of paper and on the backs of envelopes—on anything that came handy. The scrawls would be destroyed immediately in the same inconsequential manner in which they had been produced. After a time I became sufficiently interested to inquire into the nature of these sketchy objects. John showed them freely and jokingly said their underlying motif was what he called "the Lotus of the Calumet." They were the spikey shapes and bud-like things with twisted tendrils which I had begun to recognize in more sophisticated form in the ornament which the young architect, Louis Sullivan, was employing. John said that was but natural as originally he had given Louis the idea.[43]

Sullivan would go on to formulate a radical, polychromatic architecture, clad with foliate, botanically derived ornament, that today is heralded as revolutionary.[44] In dialogue with Edelmann, Simonds gained an introduction to the heady intellectual climate that informed, if not produced, it. Even if Sullivan's path did not cross Simonds's at Jenney's office, the two would converge in 1889–90, when the architect designed a pair of exquisite tombs at Graceland. Although little is known Clarence Arey's later career, he did, in fact, complete the professional

transition from civil engineering to architecture under Jenney's tutelage.[45] Irving Pond later won acclaim as one of the leading architects of the Prairie School, a nativist design movement founded on a
design appreciation of its namesake landscape type.[46] Had Simonds
remained in this environment, he probably would have achieved no
less. Ultimately, then, his decision to work for Jenney also propelled
him into the orbit of Chicago's increasingly progressive architects.
Contact with his remarkable employer and the kindred spirits he attracted no doubt affirmed Simonds's career decision. More discreetly,
Jenney's and Edelmann's esteem for the natural world demonstrated
its architectural relevance and may have further intensified his resolve
to pursue architecture.

Simonds's architectural enthusiasms, however, likely received
only antipathy, not encouragement, from Bryan Lathrop. In Lathrop's view, albeit one recorded years later, landscape gardening was
"not only one of the fine arts but . . . has possibilities of development
of which the others are absolutely incapable."[47] Indeed, for him it
was the only fine art to have progressed in the nineteenth century;
in his view "all of the others" had "distinctly retrograded." Foremost
among these, architecture, he asserted, no longer existed "as a cre-
ative art": "In the place of the mighty builders of the past we now
have schools of architecture which formulate rules based on their
work, and the best architects of our age are the most successful copy-
ists." He even went so far as to dismiss two nonhistoricist exceptions,
"'L'Art Nouveau' of Paris" and the "'Secession Style' of Vienna," as
"architectural aberrations." That Bryan Lathrop's aesthetic taste was
fundamentally historicist in nature is best confirmed by his choice of
architects to design his own house. Eschewing local innovators such
as Louis Sullivan, in 1890 he looked east and awarded his commission to the New York firm of McKim, Mead & White, noted for its
Greco-Roman neoclassicism.[48] McKim, Mead & White would later
go on to contribute several major buildings to the "White City" at
the World's Columbian Exposition, held in Chicago in 1893.

Simonds's decision to leave architecture for landscape gardening was not one reached overnight—his shift between the profes-
sions would require nearly five years, beginning when he first met
Lathrop in 1878. Despite taking up Graceland's superintendency
in 1881, he still wanted to practice architecture and continued his
partnership with Holabird. He also had contact, through John Edelmann, with the new ideas of Louis Sullivan. Graceland's burgeon-

ing work intervened, however, and within months he found himself unable to devote adequate time to the practice. Nonetheless, he did not resign. Later that same year, again drawing on Jenney's office staff, he and Holabird took on a third partner, the architect Martin Roche (1855–1927).[49] Along with the press of cemetery work, there was apparently another reason Holabird and Simonds made the new addition. Indicative of their ultimate professional ambition, the two had "decided that they were 'long on engineering and short on architecture.'"[50] Undoubtedly, Jenney would have been less than pleased to lose Roche, a seasoned office veteran of nearly a decade's standing, and that Roche was departing to join Jenney's former employees—now his rivals—must have added insult to injury.[51]

Throughout 1882, Graceland and Holabird, Simonds & Roche cross-fertilized. That spring, Simonds made a visit home to Grand Rapids, where he called on a family friend and aspiring architect, Edward Renwick (1860–1941).[52] Simonds did not find him to be "looking very well," so he offered the remedy of summer employment at Graceland. Keeping the younger man at work "out-of-doors

7.2. Graceland's Buena Avenue office around 1904; Lake Hazelmere is visible in the foreground.
From *Graceland Cemetery* (Chicago: Photographic Print Co., 1904), courtesy Trustees of the Graceland Cemetery Improvement Fund.

all the time" would be good for him, Simonds believed, but the job offer also registers the Graceland work as considerable enough at that time to require an assistant. Accepting Simonds's invitation, Renwick started work that May; it seems Graceland, not Holabird, Simonds & Roche, employed him. By the summer's end and despite Renwick's assistance, Simonds was still unable to contribute equitably to the partnership. He was not yet prepared to leave it, however. Beginning that October, he donated his assistant's services to the firm in lieu of salary. There, Renwick would go on to achieve prominence as an architect, replacing Simonds as the firm's third partner.[53]

Near the year's end, Graceland launched a new building project, its most substantial one since erecting the gateway structure some twenty years earlier. (Fig. 7.2) The cemetery now planned to construct a new office and railway station building at its eastern border, adjacent to the newly extended Chicago and Evanston Railway.[54] The project would also require landscape gardening, not only for the area around the new structure but also for a new entrance at Buena Avenue. (Fig. 7.3) When selecting an archi-

7.3. The Buena Avenue entrance around 1904.
From *Graceland Cemetery*, courtesy Trustees of the Graceland Cemetery Improvement Fund.

tect, Bryan Lathrop neither returned to the pair who had probably designed the cemetery gateway, Asher Carter and Augustus Bauer, nor did he placate Jenney with the job.[55] Instead, enlarging his patronage to bolster his landscape gardener's fledgling partnership, he engaged Holabird, Simonds & Roche.[56] Simonds was now connected to Graceland not only through landscape gardening but also through architecture. Although the new commission presented him with an immediate opportunity to practice as an architect, it is impossible to reliably distinguish which partners' hands contributed to the design. As his uncle Thomas Bryan had done earlier with Carter & Bauer, Lathrop continued to patronize Holabird, Simonds & Roche. Around this time, for instance, he also commissioned them to design furniture for Graceland's city office.[57]

At the same time, Lathrop presumably continued his campaign to persuade Simonds to abandon architecture for landscape gardening. Perhaps he hoped the revenue these new jobs garnered the firm would enable it to release more, if not all, of Simonds's time to Graceland. Cemetery work still evidently preoccupied Simonds throughout the coming months. In September 1883, he finally resigned the partnership, severing his last link to architecture.[58] Even after Simonds's exit, Lathrop would continue to involve Holabird & Roche at Graceland, whenever he was in need of an architect.[59] Simonds was now committed to a career in landscape gardening. He apparently never looked back. Entering a new profession through Graceland's portal, he soon turned the cemetery into an outdoor laboratory for his first landscape design experiments.

LATHROP'S MENTORSHIP

Near the end of his career, Simonds more than once remembered his transition from architecture to landscape gardening in print. "As work [at Graceland] progressed," he wrote in one account, "I was advised to read up on landscape gardening, a subject that was entirely new to me."[60] He consulted, among others, A. J. Downing's books *Landscape Gardening* and *Rural Essays*, as well as his *Horticulturist* articles; Humphry Repton's *Landscape Gardening*, which he described as "interesting"; and William Robinson's *Parks and Gardens of Paris*.[61] As these titles mirrored the ones Bryan Lathrop had studied earlier, it seems likely that it was he who recommended them to Simonds. Like Lathrop, Simonds would come to reject the regular geometry of formal gardens and embrace a more naturalistic approach. Simonds added that he became more interested in landscape gardening as he continued to read on the topic, and that the botanical studies he made as a youth were a great help. He found "going into the country" to select "native growth for use at Graceland" no less enjoyable. As his involvement with the cemetery accelerated, his disappointment with not practicing architecture diminished.

Ultimately, Simonds wrote, "I found that I liked the subject more than any I had studied, and this liking, together with Mr. Lathrop's influence, changed my life's work from architecture to landscaping."[62] But Simonds's own recollections of the episodes that culminated in the launching of his career as a landscape gardener must be considered critically. By the time he wrote them, decades later, his esteem for Bryan Lathrop had apparently grown to heroic proportions. This was most tellingly and poignantly registered in 1920, when Simonds dedicated the only book he ever wrote, *Land-scape-Gardening*, to "the memory of Bryan Lathrop, to whom all fine arts made a strong appeal and whose influence has been felt in each page of this volume and in all the professional work of the author."[63] Simonds even appended to the text two of Lathrop's own essays on the subject.[64]

The great esteem in which Simonds held Lathrop seems to have led him to suppress recognition of any role William Le Baron Jenney might have played in shaping his career. Even if accidental, this has diminished, if not precluded, an appreciation of Jenney's contributions not only to Simonds's professional beginnings but also to Graceland. There, for instance, Simonds credited the elder designer only as "the architect who had drawn an outline of the lake and planned its outlet." As an example of Lathrop's authoritative role in the cemetery's development, Simonds recorded an incident in which Lathrop approached him and offered a diagnosis: "You have not had experience in grading, and Mr. Jenney does not know about grading."[65] To remedy this situation Lathrop proposed that the three visit Spring Grove in Cincinnati and consult with Adolph Strauch. Whether they made this particular trip is unknown, but in any case it is inconceivable that Jenney did not "know about grading." His knowledge of grading and earthworks was, in fact, comprehensive, and this particular area of expertise was probably central to his winning the Graceland commission.

Why Lathrop—or was it Simonds?—chose to ignore or downplay this fact is unclear.

Jenney, however, was not the only one of Graceland's designers to be marginalized in Simonds's later accounts. Although he did concede Jenney's involvement, he chose to overlook William Saunders, H. W. S. Cleveland, and John A. Cole completely. In one of his cemetery retrospectives, for instance, he recognized Swain Nelson as having designed "the first, or oldest, part of the cemetery," but William Saunders is not mentioned not at all.[66] Elsewhere in the same article, Simonds identifies "Cleveland French"—clearly meaning H. W. S. Cleveland—not as his Graceland predecessor, but only as one source of Bryan Lathrop's knowledge of landscape gardening.

In another historical account, Simonds placed Lathrop's aesthetic judgments on par with those of Cleveland, among others.[67] Again he chose not to acknowledge Cleveland as one of the cemetery's earlier landscape gardeners. It is quite possible that Simonds simply forgot that Saunders was involved with Graceland in the early days; his work there was, after all, accomplished in a short period of time nearly two decades before Simonds's arrival. But Simonds could hardly have been unaware of Cleveland's contribution to the cemetery. Bryan Lathrop himself, after all, had participated in the decision to hire Cleveland, and it is difficult to believe that he would not have identified the elder landscape gardener's, or for that matter Saunders's, work to Simonds. But apparently for Simonds it was he and Lathrop, preceded only by Swain Nelson, who were responsible for Graceland's design, a view still popular today.

When he reflected in print on how he got started in landscape gardening, Simonds, most curiously, did not actually name the person who first advised him to "read up on landscape gardening." His decision to take up this suggestion had proven to be life-changing, setting him down on a new professional path that led eventually to success. He could hardly have forgotten the source of this advice, yet he chose not to reveal it. As we've seen, in his reminiscences Simonds hinted only that his reading, along with Lathrop's influence, had prompted his decision to become a landscape gardener.[68] This juxtaposition suggests that it was at Lathrop's prompting that he began reading on the topic. Yet Simonds routinely credited his patron's influence—in the dedication to *Landscape-Gardening* and elsewhere—and that he did not name Lathrop here seems odd. Could it be that his resignation from Jenney's practice had actually ruptured the relationship between the two men, a circumstance exacerbated when William Holabird and later Martin Roche followed suit? If so, then Simonds's silence on this matter was possibly a deliberate means of concealing the fact that it was Jenney, not Lathrop, who first steered him toward this new pursuit.

It must be remembered, of course, that Simonds wrote his accounts of Graceland's history in the final years of his career, and that by then his memory of events that took place nearly half a century earlier might well have dimmed. Nevertheless, Simonds's repeated public acknowledgments of Lathrop's formative influence, although they were clearly sincere, were also perhaps motivated by ambition to entrench his mentor within the annals of landscape gardening. Thus we cannot completely dismiss the possibility that Simonds's backward glances were made through a revisionist lens.

In 1881, with Lathrop's encouragement and guidance, Graceland's new superintendent furthered his hands-on practice in his newly adopted profession. His protégé later remembered Lathrop as a knowledgeable and inspiring teacher: "To be with him at his country home in Elmhurst, to travel through the country with him, and, above all, to be with him at Graceland, would give one an education in what he always called 'landscape gardening.'"[69] Indeed, writing after more than a half-century of professional practice, Simonds still ranked Lathrop's aesthetic judgment "as high as that of Repton, Robinson, Downing, Olmsted, Cleveland, or Strauch." Lathrop had probably begun mentoring his new landscape gardening recruit when Jenney first assigned him to Graceland, three years earlier. Now, with Jenney no longer involved at the cemetery, Lathrop no doubt amplified his instruction, likely grooming the new superintendent in accordance with the ideals he later outlined in an essay Simonds printed at the end of *Landscape-Gardening:*

> The ideal landscape-gardener should have a vast range of knowledge. He must be a botanist, and he must know the nature, the habits of growth, of trees, shrubs, and plants, and those which are adapted to each region; he should know the chemistry of horticulture, and the nature of soils; he should be an engineer, as the basis of his work is the grading and shaping of the earth's surface; he should have a knowledge of architecture, as his work will often make or mar the work of the architect; and finally he must be an artist to the tips

of his fingers; the more artistic he is the better landscape-gardener he will be.[70]

By now Simonds, through his civil engineering education and his three years in Jenney's office and on site at Graceland, had likely achieved fundamental technical proficiency as a landscape gardener. Measuring him against the full spectrum of Lathrop's criteria, however, he possibly had yet to satisfy his employer's ambition that he become "an artist to the tips of his fingers." Cultivating his landscape gardener's aesthetic design sensibilities was probably next on Lathrop's agenda. Presumably early in Simonds's tenure as superintendent, for instance, he and Lathrop toured Graceland's counterparts in "Cincinnati, Cleveland, Buffalo, New York, Boston, Philadelphia and other cities," studying their "grades, planting and other features."[71] Simonds recollected in detail a visit he and Lathrop made to Cincinnati's Spring Grove, where they encountered Adolph Strauch, for him "the fountain head of the modern cemetery." There, the pair found him "down in a ditch laying drain pipe," and "as he climbed out," Strauch explained that his workman could not otherwise properly complete the task. Strauch then guided the Graceland contingent through Spring Grove's "principal features": the "rolling lawn," "groups of trees and shrubs," and "drives with graceful curves and low turf margins which he said should always be tangent to the surface of the roadway." In order to demonstrate and more fully convey his picturesque landscape effects, Strauch "emphasized and framed the pictures with a little mirror" as the trio walked. Simonds mentions unnamed "cemeteries at Cleveland and Buffalo" as among those that displayed Strauch's influence. Indeed, "everywhere, even in the east," he concluded, "it was conceded that Spring Grove was the most beautiful cemetery in the world."

Of all the cemeteries Simonds and Lathrop studied, Spring Grove would prove to be the most important for Graceland, as it was for many rural cemeteries. We must keep in mind that the landscape of Graceland had already registered the influence of Spring Grove before Lathrop and Simonds began their study. It was H. W. S. Cleveland who first shifted the focus of Graceland's design away from the pioneering eastern rural cemeteries—such as Mount Auburn and Laurel Hill—westward to Ohio and the lawn plan ideals Strauch applied at Spring Grove. Whether or not Cleveland's lawn plan variant was ever implemented, Lathrop and Simonds still seemed compelled to seek out its original source.

LATHROP AND SIMONDS'S WORK
ON THE EASTERN AREA

Now Lathrop and Simonds were to concentrate their activities in the eastern half of the cemetery, an area of "about fifty acres—nearly half a mile north and south—and of varying widths."[72] What progress had already been made there by 1881? Simonds notes that Lake Willowmere and its surrounding earthworks, begun near the end of 1878, were finished by the end of the next year.[73] Beyond this, the project's status then is unclear, but an overview of Simonds's career published in 1931 supplies us with an important clue; it reports that his superintendency began "after the engineering part attached to Graceland Cemetery had been accomplished."[74] Presumably "engineering part" meant the earthworks and the construction of drives and pathways. Thus it is likely that the drainage and grading activities, such as excavating two additional water bodies and laying out the network of thoroughfares, had consumed 1880 and much of 1881, but by now were complete. Given that the civil engineering operations required to realize Jenney's vision were indeed extensive, this scenario seems likely enough. According to one historical account, apparently written by Simonds's successor, H. J. Reich, the cemetery's entire eastern half was elevated with "thousands of yards of material" excavated for the lakes and its soil improved with "many thousands of loads of rich clay soil" hauled from land near the north branch of the Chicago River.[75] This was a very considerable undertaking that could easily have required a year or more to complete, remembering too that the frozen winter months would have impeded or possibly even halted progress. There were likely still finishing touches yet to be made on Jenney's artificial terrain, however; around this time, for instance, the roads, "at first only sand, were gradually improved by adding gravel, clay and gravel, slag and finally broken stone and asphalt or tar to form a hard smooth surface."[76] By the time Simonds became superintendent, then, the last large task to be accomplished at the cemetery was probably to embellish Jenney's sculpted landforms with plantings.

The minutes of the board of managers' meetings have already told us that Jenney's plans called for the addition of large trees. As his earlier work at the West Parks demonstrates, the landscape gardener was indeed an accomplished plantsman. The extent to which he had developed Graceland's planting scheme before departing the project is unknown; his plan may have been only a notional or schematic one, rather than comprehensive and detailed. Reich says that none of the designs by Simonds and Lathrop's predecessors "included any plan for the planting."[77] Lathrop, apparently confident of his horticultural abilities, perhaps saw Jenney's role as more technical than aesthetic. Now that the lakes, roads, and earthworks were essentially finished, Lathrop's own design ambitions, it seems, led him to use the "discretion as to the number and cost of large trees to be planted in Graceland" that the board had reserved earlier to set aside Jenney's planting scheme, whatever its state. Graceland's planting, like its engineering works, would prove a major undertaking, requiring not years but decades to achieve and perfect.

Before planting, however, the new sections required subdivision into individual burial plots. Having already revised H. W. S. Cleveland's Section O with large plots around 1878, Lathrop now instructed Simonds to lay out another one, this time "with still larger lots, each having a frontage of 100 feet."[78] Although Simonds remembered the project's site as a "new" section, the map made by Charles Rascher around 1878 shows that what the pair were resubdividing was actually another Cleveland section, V. By the time they reapportioned it, this section had been renamed Ridgeland and its outline revised by Jenney. Comparing Rascher's map to the 1884 *Map of Graceland Cemetery, Cook Co., Ill.* also shows that Lathrop and Simonds introduced these new, larger plots elsewhere in the cemetery. In the spring of 1882, for instance, Edward Renwick was at work "as rodman for [Simonds's] surveying," making "calculations to determine the area of the various lots."[79] Although Renwick did not identify which section he surveyed, it was probably Lakeside, the first new section to be subdivided.[80] Lathrop and Simonds did not restrict their activities to Graceland's new sections. In fact, the 1884 *Map of Graceland Cemetery, Cook Co., Ill.*, according to part of its legend, was published to document four of their resubdivisions within the cemetery's older, western half.[81] By 1884, they had already modified portions of Sections M, N, and P, all designed earlier by Cleveland. In Section N, the pair laid out a new pathway called Evergreen Path and lined it with comparatively larger plots. They similarly revised a portion of Section P, apparently an expansion, opposite the Ridgeland section. The lots along the borders of

Section M, already relatively large, were now, rather unusually and for unknown reasons, resubdivided into smaller dimensions. Most remarkably, Lathrop and Simonds also resubdivided the core of Section G. There, somewhat surprisingly, they sacrificed the cemetery's grotto retreat, one of the main features of William Saunders's original plan, to gain burial space, subdividing its peripheral parkland into lots and creating a new "vault section" at its center.

Once subdivision was complete, markers within the individual plots were the next concern. As Simonds later reported, Lathrop believed that "tombstones and other stone work marred instead of enhanced" beauty.[82] Apparently through his persuasiveness with the board, "headstones, at first unlimited as to height, afterwards limited to thirty inches, were subsequently reduced, in accordance with rules adopted by the board of managers, to eight inches, four inches, and finally even with the lawn." Next up in scale were monuments, and they too received Lathrop's critical scrutiny. "It was universally recognized," Simonds conceded, "that monuments belonged in a cemetery," but likely through Lathrop's influence Graceland made a rule "restricting them to one for each lot," since "it was thought by those in charge that for a given sum of money a much better effect could be secured with a larger lot and a small monument than with a large monument and a small lot." "One section," he recounted, "was reserved for those who did not care to have monuments, and portions of other sections have been thus restricted."[83]

In 1883 the guidebook *Chicago Illustrated* observed that the intention of Graceland's management was "to preserve the wide and beautiful sweep of the lawns by excluding, as far as possible, stone and marble from the new grounds, the monuments being restricted in number, and the headstones being kept low and unobstructive, and all the old-fashioned and repulsive stone edges, fences, posts and chains, and all the other unsightly lot-enclosures once in vogue being forbidden."[84] With the changes put in place by Lathrop, the adaptation and codification of Strauch's ideals at Graceland was now apparently complete.

A NEW ENTRANCE AND THOROUGHFARE

Also around this time, Graceland's decision in 1882 to erect its own railroad station and office resulted in new landscaping. Today long demolished, the building originally stood at the cemetery's eastern limits, near the south end of Lake Hazelmere. Shortly after it was completed, Simonds would have been responsible for planting the station's grounds to merge them within the cemetery's wider landscape. But another aspect of this project also entailed designing a new landscape. The new railroad terminus meant that another entrance would be created, aligned near the intersection of Stella Street and Buena Avenue at the eastern boundary. Visitors arriving by train would now enter Graceland by strolling down the new Tanglewood Path. Designed by Simonds, likely in consultation with Lathrop, this thoroughfare entry originated just inside the new Buena Avenue entrance, at the intersection of Jenney's Broad and Wildwood avenues. From there it arced southwest, spanning the Edgewood and Bellevue sections. After crossing Main Avenue, Tanglewood continued until it terminated at Greenwood Avenue. Significantly, constructing the new path would also have entailed reengineering the earthworks in the burial sections it passed through. Accentuated with trees and shrubs along its margins, Tanglewood was not merely a path, but more a spatial corridor or linear parkland, somewhat reminiscent of a country road. In the coming years a handsome chapel would be built near the end of Tanglewood Path.

PLANTING

Our knowledge of the history of Graceland's planting is fragmentary at best. Unlike stone, brick, and steel, vegetation, landscape gardening's principal medium, is dynamic and sometimes ephemeral. Indeed, the cemetery's arboreal mantle today is much diminished from the one Lathrop and Simonds cultivated. We must turn to textual sources, a handful of drawings, and contemporary photographs in an effort to recover it at least partially. Neither Lathrop nor Simonds is thought to have ever made an overall, comprehensive plan of Graceland's planting. It seems doubtful that Lathrop would have required Simonds to commit his ideas to paper in advance for his approval. Instead, the pair apparently conducted their design experiments in the cemetery itself, rather than on the drafting board; like Adolph Strauch at Spring Grove, they worked largely without the aid of drawings. Fortunately there are several notable exceptions; a planting plan for the eastern half of the Ridgeland section is the most important of these. (Fig. 7.4) Simonds also made detailed planting plans for a num-

7.4. O. C. Simonds's planting plan for the Ridgeland section, noted for its expansive burial plots.
Courtesy Chicago History Museum (ICHi-61230).

ber of individual burial plots, presumably so they could be approved by the owners of these lots before work started. (Figs. 7.5–7.7) Only a very few of these drawings survive, however.

But Simonds wrote prolifically on planting design subjects, and some of his articles provide details about his work at Graceland. For example, he left a firsthand account of his initial planting activities at the cemetery. Recalling that Jenney's newly made terrain was then barren and treeless, he writes that he was quickly in need of trees and shrubs, and he traveled to the country to select "native growth" from outlying farms.[85] Robert Leesley, Simonds's predecessor as superintendent, followed afterward to buy the native trees and shrubs the landscape gardener had identified and to deliver them to the cemetery. We've seen that the board's original discussion in January 1880 included mention of large trees as part of Jenney's design, and Simonds recollected, for instance, securing elms and other trees that were "fourteen, sixteen and eighteen inches in diameter." With such specimens Lathrop and Simonds were apparently hoping to achieve immediate spatial and visual effects. Along with trees, the cemetery also purchased shrubs by the wagonload— "all a team could haul."[86]

Lathrop contributed actively to the planting designs. At one

(Fig. 7.5, above; 7.6, opposite; 7.7, page 156) Planting designs for individual plots. These drawings were probably made to secure the respective owners' approval in advance of implementation.
Courtesy Chicago History Museum (ICHi-61226, ICHi-61223, ICHi-61224, ICHi-61228).

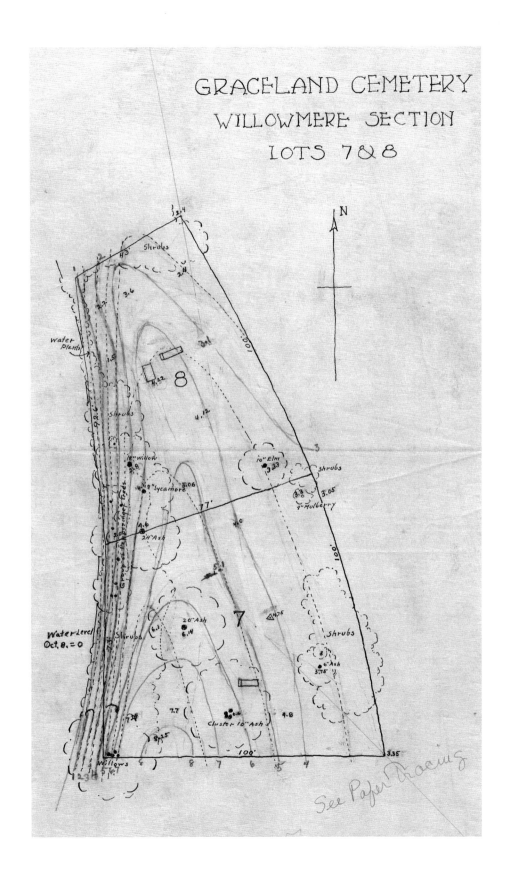

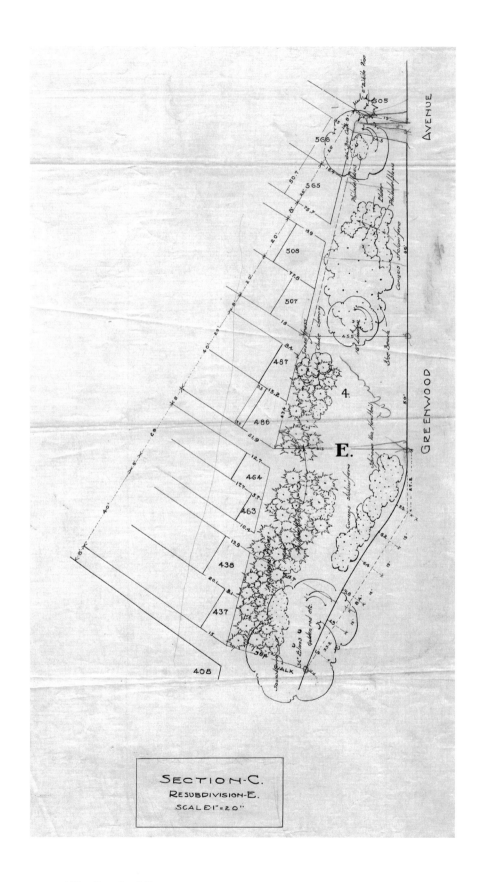

SECTION·C.
RESUBDIVISION·E.
SCALE: 1"=20"

point he became distressed that it was possible to "see from one end of the cemetery to the other," Simonds recalled, and directed his landscape gardener to "get more shrubbery" and compose it so that parts of the cemetery were hidden.[87] (Figs. 7.8 and 7.9) Shrubbery, Simonds noted, "could be used in concealing the boundaries and separating one part of the ground from another, as well as for its own charming appearance in foliage, flower, and fruit."[88] Lathrop would also make what Simonds called "equally suggestive and appropriate remarks about other features of the work."[89]

What criteria informed Lathrop and Simonds's selection of tree and shrub species? Simonds noted that "weeping willows and evergreens, especially the tall specimens, narrow evergreens like arbor vitae and junipers had formerly been considered the trees

7.8. View across Lake Willowmere, probably around the 1880s. Courtesy Chicago History Museum (ICHi-i61406).

most appropriate for cemeteries."[90] Indeed, the Lake Willowmere
and Evergreen sections of Graceland continue this tradition.
Although Simonds is celebrated today as an early champion of
native species in his work at Graceland, he and Lathrop apparently were not purists. "Mr. Lathrop," Simonds reported, "advocated the introduction of all plants that are beautiful and thought
the cemetery should become an arboretum in which every hardy
plant might be found."[91] In 1931, after more than a half-century's
work at Graceland, Simonds recommended that native shrubs be
supplemented with "all the so-called old-fashioned shrubs such
as lilacs, syringas, honey-suckles, barberries and snowberries
and also the newer kinds."[92] "And when it comes to flowers," he
instructed, "use all those found in our grandmothers' gardens,"

in conjunction with "our native flowers." "Rather than to supply a [species] list," he explained, he aimed to "suggest the greater use of indigenous plants." One should note that Simonds here refers to a greater, not an exclusive, use of native plants. Hardiness, rather than a regionalist aesthetic impulse, appears to have been the paramount—and economically sensible—criterion for species selection.

Simonds began enhancing the plantings within the limits of the cemetery's original layout and artfully transformed its new and undeveloped portions. As Walter Creese writes, he fused the "initial rural cemetery type of Mount Auburn" with Adolph Strauch's lawn plan as it was expressed at Spring Grove.[93] Modifying Strauch's greensward technique, he masterfully created a sequence of spatial vignettes. These outdoor pictures, although implicitly indebted to English picturesque naturalism, also responded to indigenous landscape cues. The level prairies of the Midwest varied greatly from the more sylvan, rolling terrain of Massachusetts or even Ohio. The vast dome of the prairie sky and its omnipresent horizon, together with the openness and immense lateral spread of Graceland's setting, demanded much from plantings. To meet these challenges, Simonds employed an internal axis—defined and accentuated by foreground plantings of native trees and shrubs—as a design device to establish and visually control a sense of infinite space within the cemetery. Simonds's axial constructions reduced the boundless prairie to a series of controllable, comprehensible pictures, enabling the viewer to establish his psychological place on the land through the carefully determined foreground. Such "long views," as they were later labeled, were Lathrop and Simonds's means "to keep in touch aesthetically or spiritually with the smoky line of the distant prairie horizon, often expressed in cream or pearl grey tonalities, just where it joined the sky."[94] (Figs. 7.10 and 7.11) At Graceland, this axial extension of physical distance implied a more eternal life. This illusory effect depended somewhat on maintaining a greensward free of markers and monuments—a goal that would prove elusive.

In 1882 Bryan Lathrop wrote in an advertisement for the cemetery that the landscape features of the new area included "a beautifully undulating surface; picturesque lakes; fine lawns; smooth roads; [and]

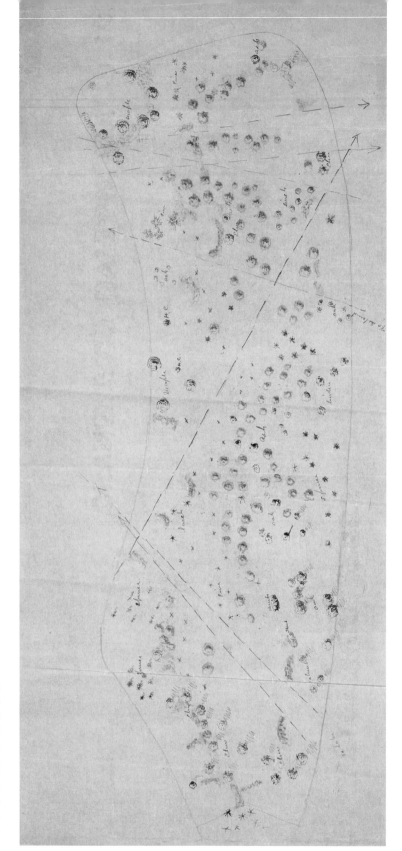

7.10. Planting plan
of the Lakeside
section; arrows
delineate spatial
corridors within the
trees and shrubs,
later termed by
Wilhelm Miller
"long views."
Courtesy Chicago History
Museum (ICHi-61221).

a great variety of trees and shrubbery."[95] These attributes presumably were ones already realized, not projections, and he must have been pleased with them. Indeed, during Simonds's tenure Graceland became a refuge from the city, offering peace and the gentility, civility, and dignity that were hard to come by in the ever-growing city of Chicago.

7.11. A "long view" in Graceland.
Courtesy Chicago History Museum (ICHi-61405).

Epilogue

Graceland's ongoing development was not limited to planting. By the mid-1880s a network of drives and paths, an entry gate structure, and a railroad station had been constructed. Around this time, the cemetery's landscape infrastructure gained a new addition: an extensive irrigation system. A steam pump directed water, drawn from a natural spring, through a network of underground pipes to irrigate the lawns and plantings and also to feed lakes Willowmere and Hazelmere and Lotus Pond. "Thus," a contemporary history of the area wrote, Graceland's "wide and beautiful lawns are always cool, sparkling and green," even during "the most parching weather."[1]

In 1885 the Chicago and Evanston Railway at last connected Graceland to the city and the cemetery's rail station became functional. The convenience of the new rail link would soon prove popular and likely informed the cemetery's decision to erect a chapel to accommodate onsite memorial services and a mortuary crypt. In 1888 Bryan Lathrop again turned to Holabird & Roche, the firm that had designed the railway station six years earlier; now, as Robert Bruegmann notes, the partnership had "started to expand rapidly, taking on dozens of new draftsmen."[2] Holabird & Roche responded with what Bruegmann describes as a "solid, almost primeval granite-walled" structure.[3] (Fig. 8.1) Up until then, as we have seen, the Gothic Revival gateway spanning the cemetery's carriage or main entry at Clark Street also in-

8.1. Graceland's chapel and mortuary crypt in 1904.

corporated a small chapel. The new chapel was a sweeping design departure from this now stylistically archaic structure. It also markedly contrasted with the cottagelike Graceland railroad station. With this chapel Holabird & Roche established a new architectural aesthetic for the cemetery. Bruegmann's description of the chapel as "almost primeval" is especially apt, as the firm's design took its cues from the massive stone structures of the Bostonian architect Henry Hobson Richardson. Drawing inspiration from his close friend and collaborator Frederick Law Olmsted, Richardson sought to imbue his buildings with a sense of antiquity or the primordial, relying, for instance, on rough-hewn stone construction.[4] Richardson's John J. Glessner house in Chicago (1885) offered an inspirational source close to home.

At the outset of the project Holabird & Roche, presumably in consultation with Lathrop and Simonds, identified an undeveloped portion of Section C, southwest of Greenwood Avenue, as the best position for the new building. The chapel is comparatively far removed from the main carriage entry on Clark Street, and this may simply reflect that the Buena Avenue entrance had now become more popular. But Simonds would later advocate that a cemetery chapel "be placed some distance within the grounds to give it greater seclusion and quietness," and it may be that he made the same argument to Lathrop about the siting of the Graceland structure.[5] Constructed of Waupaca granite from Wisconsin, the chapel appears to emerge from the undulating topography. The building complemented Graceland's landscape chromatically as well, suffusing the scene with the earthy red, brown, and green tonalities of its granite walls and reddish-brown Spanish tile roof.[6] Here, architecture and landscape merge and become continuous. One critic, reviewing the firm's work in 1897, also discerned this fusion, commenting that the chapel had become "almost covered with climbing vines."[7] Simonds adorned the chapel grounds with trees and shrubs, and Tanglewood Path now gained additional prominence as the direct pedestrian link between the new building and the railway entrance at Buena Avenue. In order to enable carriage access to the chapel, Simonds projected a new drive—roughly a half cul-de-sac, oblong in shape—off of the southwest side of Greenwood Avenue. Within the chapel itself Simonds was able to indulge in a bit of landscape gardening as well. One lower section of the roof had a large skylight, and beneath it grew a "semi-circle of palms and potted plants," gracing the chapel interior with an ethereal profusion of tropical greenery, even in the otherwise bleak winter months.[8] (Fig. 8.2)

8.2. The chapel's interior in 1904.
From *Graceland Cemetery,* courtesy Trustees of the Graceland Cemetery Improvement Fund.

Although its landscape structure or frame was likely finished, new monuments would continue to be added to those already in place. Virtually from its opening, the cemetery began to accumulate markers and other monuments, and by the last decades of the nineteenth century it was studded with hundreds of monuments of varying size and aesthetic merit. (Fig. 8.3) In terms of style they ranged from Egyptian Revival columns, obelisks, pyramids, and mausoleums to neoclassical columns and temples. Interspersed among them are crosses large and small, as well as more diminutive sculptural markers. Now the cemetery would begin to gain monuments designed by architects of national repute. Of these, Louis Sullivan's monolithic tombs for the lumber and real estate magnate Martin Ryerson (1889) and for Carrie Eliza Getty (1890), the wife of another lumber merchant, Henry Harrison Getty, are the most remarkable. Grand monuments were not compatible with the "lawn plan" that Simonds and Lathrop adopted, but, as we saw in the previous chapter, Simonds and Lathrop decided to permit them as they realized that it was "universally recognized that monuments belonged in a cemetery."[9]

8.3. Victorian-era grave markers at Graceland (c. 1880). Courtesy Chicago History Museum (ICHi-61235).

At first glance, Ryerson's monolithic granite monument appears to be an isolated object, an impression amplified by fact that its model—an ancient Egyptian mastaba, a mud-brick tomb with flared sides, capped with a pyramid—is remote in both space and time. (Fig. 8.4) But Sullivan actually designed this monument to reflect—in a literal sense—the wider setting. "On a clear day," Sullivan's biographer Hugh Morrison astutely observes, "the polished black walls form a dark mirror in which one sees ethereal reflections of green trees, blue sky, and moving clouds."[10] The Getty tomb is positioned near Willowmere, the artificial lake designed by Sullivan's former employer, William Le Baron Jenney. (Fig. 8.5) Like the Ryerson monument, the Getty tomb is in a dynamic relationship with its context. "When the shadows from the trees fall across [the tomb's] surfaces," Walter Creese writes, "the two ingredients of Sullivan's ornament—underlying, powerful geometry, and rich,

8.4. Louis Sullivan's monolithic tomb for Martin Ryerson (1889).
Photograph by Carol Betsch, 2009.

overflowing, arabesque in natural patterns—are merged and revitalized."[11] Collectively, according to Morrison, Sullivan's tombs, unlike more conventional monuments, "celebrate, not the permanence of death, but the permanence of life; they express in terms of lyric beauty that a man or a woman has *lived*, not merely that he or she has died."[12] Simonds himself apparently esteemed Sullivan's mausoleums, despite their incompatibility with the lawn plan ideal. In his book *Landscape-Gardening*, he reproduced an image of the Getty tomb under the caption "The nobility of trees and background," describing it as "a tomb with a satisfactory setting, and simple in all its details."[13]

CREMATORY AND OTHER NEW BUILDINGS

In 1893 Graceland's recently built chapel and crypt gained a new function with the construction of a crematory—Chicago's first—beneath it.[14] The chief reasons for the facility, Simonds told the *Chicago Tribune* a few years later, were to satisfy "those who desired to avoid expensive funerals and interments" and to provide "a safe and efficacious manner of disposing of the bodies of those who were victims of contagious diseases."[15] Sanitary concerns may in fact have been the prime motivating factor; in an article written in 1930, Simonds recalled that when he was returning to New York from Europe in 1892, his ship was quarantined for suspicion of cholera, and with the World's Columbian Exposition set to open in Chicago the next year, Bryan Lathrop anticipated the very real possibility of an epidemic and consequently commissioned a crematory for Graceland.[16] Once the crematory was in operation, the question soon arose of the "most satisfactory way to dispose of the receptacle containing the ashes," and Simonds believed that it should, like a coffin, be buried in a family plot, which would preserve the "sentiment connected with trees and shrubbery, the songs of birds, and restful landscapes."[17]

In 1895 Lathrop launched Graceland's most extensive building project up until then: a new entrance gate, administration office, and waiting room at the cemetery's original Clark Street entry. (Figs. 8.6 and 8.7) By then the North Chicago City Railway had at last connected Graceland to the city with cable cars, which likely increased the number of visitors arriving at the original entrance and may have prompted Lathrop's plan. Again choosing Holabird & Roche as his architects, he asked them to design the new

(*Opposite*)
8.5. Sullivan's monument for Carrie Eliza Getty (1890); the edge of Lake Willowmere is visible in the foreground. Photograph by Carol Betsch, 2009.

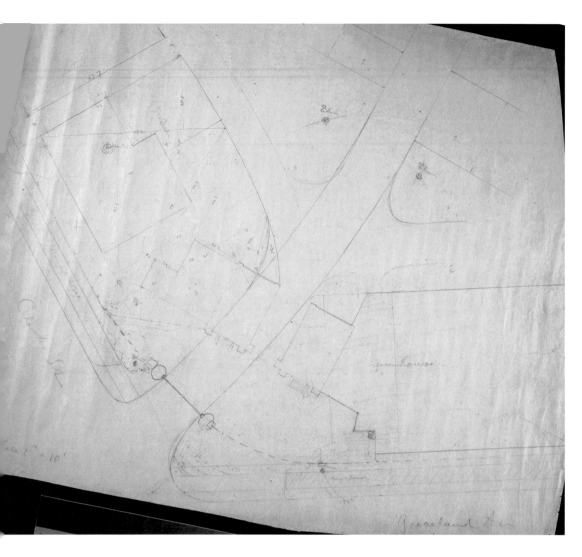

gate and buildings, as well as a perimeter fence for the cemetery.[18]
At the same time, the entry road itself was to be widened. This
sizable undertaking required the demolition of Thomas Bryan's
original combined office and gateway, erected nearly forty years
earlier, which was now antiquated and functionally outmoded. In
their array of new Romanesque designs for the project, Holabird &
Roche again employed rough-hewn Waupaca granite and Spanish
tiled roofs. Compact and low in profile, the buildings, together with
their surrounds, created, as Simonds later wrote, "a pleasing combination of architecture and planting."[19] Ultimately, as with the
earlier chapel, Graceland's broader landscape visually absorbed
the new picturesque entry ensemble; architecture and landscape
again fused into a harmonious whole.

GRACELAND IN 1900

In 1900 Paris hosted a world's fair, the Exposition Universelle, to commemorate the artistic and technological achievements of the previous century. Likely hoping to attract additional prestige, Graceland submitted a set of photographs of its grounds for display in the U.S. horticultural exhibit, and it was awarded a silver medal (a gold medal went to Spring Grove Cemetery in Cincinnati).[20] Graceland's decision to participate in the Paris exhibition may have signaled that the cemetery's landscape development was complete, or nearly so, by around 1900. That Simonds had resigned his post as superintendent in 1898, after twenty years' work at the cemetery, suggests this was the case, since a virtually finished landscape would have required less attention from its designer than a still evolving one.[21] Simonds continued as Graceland's landscape gardener and was reappointed in this diminished capacity in 1903.[22] This time, however, he was now "at liberty to practice his profession outside of the cemetery," and he soon established the firm of O. C. Simonds & Company to do so.[23] Graceland permitted Simonds to run his private practice from its railroad station office. From around 1900, then, it appears that

8.7. Graceland's new entry gate, waiting room, and office, completed around 1896. The figure standing at the entry is possibly O. C. Simonds. Courtesy Chicago History Museum (ICHi-61236).

Simonds's contributions to Graceland's overall design were, like its landscape, essentially complete.

ADDITIONAL MONUMENTS

In the new century monuments continued to proliferate, and today Graceland is adorned with dozens of historically significant monuments designed by notable architects. Although often beautiful as sculptural objects or as miniaturized architecture, many of these have only an incidental relationship with the cemetery's wider landscape. But there are some remarkable exceptions. In 1906 the sculptor Daniel Chester French and the architect Henry Bacon collaborated to produce a monument to the department-store magnate Marshall Field, set within a formal garden. (Fig. 8.8) French's allegorical figure

8.8. Marshall Field's formal garden memorial (1906). Photograph by Arthur G. Eldredge.

8.9. Potter Palmer monument.
Photograph by Rachelle Bowden (rachelleb.com).

Memory is seated atop a pedestal designed by Bacon. From this perch she gazes melancholically across the placid waters of a rectangular reflecting pool—with benches for contemplation at either end—also from the architect's hand. French and Bacon would later collaborate to design their most famous work, the Lincoln Memorial in Washington, D.C. The monument to the merchant and hotelier Potter Palmer and his wife, Bertha, built in 1921, is perhaps the most grandiose to be found at Graceland. (Fig. 8.9) Designed by McKim, Mead & White (the same New York architects Bryan Lathrop commissioned to design his own house), this Grecian temple sheltering the couple's sarcophagi is located in an elevated, commanding position overlooking Lake Willowmere.

In 1912 the cremated remains of Chicago's, if not the nation's, preeminent urbanist, Daniel Hudson Burnham (1846–1912), were interred at Graceland. Burnham is most often remembered today as the chief orchestrator of the World's Columbian Exposition (1893) and for his 1909 *Plan of Chicago* (with Edwin H. Bennett). He is also known for this legendary dictum:

> Make no little plans. They have no magic to stir men's blood and probably themselves will not be realized. Make big plans; aim high in hope and work, remembering that

a noble, logical diagram once recorded will never die, but long after we are gone will be a living thing, asserting itself with ever-growing insistency. Remember that our sons and grandsons are going to do things that would stagger us. Let your watchword be order and your beacon beauty. Think big.[24]

Burnham died in Germany, during a European tour with his family; he was cremated there and his ashes were returned to Chicago. The site selected for the burial was a dramatic one, created by his former employer, Jenney—the island in Lake Willowmere. We do not know who chose the site, but it is doubtful that Bryan Lathrop or Jenney ever intended the island to accommodate burials. In what is apparently the earliest known photograph of the island, taken around 1880, it is profusely and luxuriantly planted and adorned by a gazebo at its south end. The gazebo would have required a bridge to access it, although no drawings or photographs of one have come to light. More broadly, the presence of a gazebo suggests that the island was originally intended as a picturesque retreat not for the dead but for the living. Nonetheless, the island was now put to new use and Burnham's remains were marked with a granite boulder. (Fig. 8.10) A new bridge, called the Log Bridge, was designed to access the island from its northern end. A drawing of it survives, prepared by Simonds's employee Frank Button (1866–1938).[25] (Fig. 8.11) Whether

8.10. Daniel Burnham's simple monument on the island in Lake Willowmere (1912).
Photograph by Rachelle Bowden (rachelleb.com).

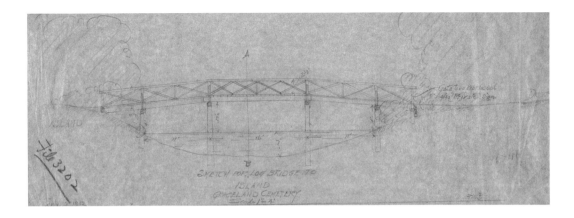

the design Button recorded was his own or Simonds's is unclear, but the drawing documents that the "gate [at the bridge entry was] to be kept locked" and marked with a "'Private' sign." As these measures indicate, the island was no longer quasi-public parkland.

At the other end of Lake Willowmere, the Goodman mausoleum, erected by lumber baron William Goodman for his son, the playwright Kenneth Sawyer Goodman (after whom Chicago's Goodman Theatre is named), is another monument whose design is informed by a marked concern for landscape integration. (Fig. 8.12) Designed by the Chicago architect Howard Van Doren Shaw in 1919, the neoclassical construction is built into Lake Willowmere's embankments. It features a dramatic waterside entry, suggesting a boat approach. The roof functions as a terrace, offering views across the lake to the island.[26]

Even beautiful monuments, however, clashed with the lawn plan Lathrop and Simonds had adopted from Strauch's Spring Grove. Walter Creese discerns a palpable tension between the pair's design ideals and Graceland's reality. "The necessity of submitting to death, as opposed to the collective urge of Chicagoans to keep forever going," he writes, "caused Graceland to seem a constant paradox."[27] In Creese's view the desire for immortality was expressed through large family monuments, such as the Potter Palmer temple. Creese concludes: "As Lathrop and Simonds's ideal was one of a vast landscape painting, so the families' was one of a well-filled sculpture gallery. The Chicago families ached for a literal, consolidated permanence, Lathrop and Simonds for a permanent beauty seen at a distance and in perspective, somewhat hazily."[28] The "monument" Bryan Lathrop selected for his father's grave poignantly accentuates Creese's point. When Jedediah Lath-

8.11. Elevation of design for a log bridge to the Willowmere island (1912), enabling access to Burnham's monument. Courtesy Chicago History Museum (ICHi-52330a).

8.12. Playwright Kenneth Sawyer Goodman's lakeside mausoleum (1919). Photograph © Kate Corcoran; used with permission.

rop died in 1889, his son arranged for an American elm nearly three stories tall to be uprooted from a Michigan farm, transported to Graceland, and replanted on his father's plot.[29]

By the opening decades of the twentieth century Graceland was no longer a rural cemetery. Now its environs were urban, annexed by the city of Chicago in 1889, and the Graceland Cemetery Company itself furthered the transformation, subdividing its properties beyond the cemetery's limits in accordance with layouts by O. C. Simonds. (Figs. 8.13 and 8.14) In 1891, for instance, the company sold a parcel of its land just north of the cemetery, marketing it as the Sheridan Drive Subdivision. It was fully developed by 1929, and today it is known as Sheridan Park. A designated historic district, Sheridan Park is comprised largely of houses and apartment hotels. In another example of the company's real estate development pursuits, Simonds prepared plans in 1914 to subdivide a parcel of land opposite the cemetery, fronting Southport Avenue. Three-story brick apartment buildings replaced the open land and remain today.

ACCOLADES

After the 1900 Exposition Universelle, Simonds's work at Graceland continued to attract praise. In 1903 it captured the attention of the New York-based horticultural writer and critic Wilhelm Miller (1869–1937), who described Graceland (and Spring Grove) as the "most perfect expression" of the "modern" or "park-like" cemetery, in the popular magazine *Country Life in America,* and he credited Simonds with making Graceland "the admiration of the world."[30] Following Miller's article was one by Simonds himself, "The Planning and Administration of a Landscape Cemetery." The pair of essays was illustrated by six photographs of Graceland. This would not be the last time Wilhelm Miller promoted Simonds and his work at the cemetery.

In 1910 Frank A. Waugh (1869–1943), a professor of landscape gardening at the Massachusetts Agricultural College (now the University of Massachusetts Amherst), called the cemetery an "American Masterpiece" in his book *The Landscape Beautiful.* That Simonds had employed what Waugh labeled the "natural style of gardening" at the cemetery was central to winning his esteem. Although he believed Simonds to be "a highly independent worker," Waugh contended that he had "still been influenced to a considerable extent by the work of the elder Olmsted." "Nevertheless," he continued, "Graceland Cemetery is peculiarly his own enterprise." Indeed, "in its present form," Waugh writes, "he may be said to have established it," and "there is hardly a piece of work to be found anywhere in the United States which is more directly and completely the personal product of one man's labors." Waugh distinguished Graceland from its more conventional counterparts:

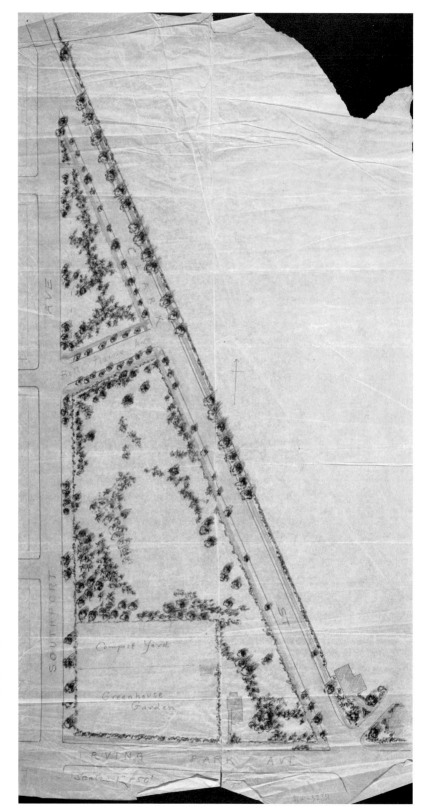

8.13. Plan of Graceland's property adjoining the cemetery, depicting its gardener's cottage, greenhouse, compost yard, and parklands (c. 1912).

8.14. O. C. Simonds's plan to subdivide the same area, here named "Southport–Irving Park Subdivision," in preparation for sale (1914).

Courtesy Chicago History Museum (ICHi-52332).

The eastern section of the present cemetery . . . has been kept practically clean of architecture, stone masonry and other mortuary gewgaws, and also of canna beds, coleus borders and the other usual vulgarities. It contains a broad, quiet stretch of lake, heavily bordered by luxuriant plantings of shrubbery and comfortable trees. Here and there are quiet stretches of unbroken lawn. From many points there develop strongly composed pictures of quiet, restful, rural scenery. The feeling of peace, quietude and rest which ought to characterize a cemetery is here realized as fully as the art of landscape gardening can realize it.[31]

In 1915, Wilhelm Miller again wrote about Graceland in *The Prairie Spirit in Landscape Gardening,* using several photographs of the cemetery to illustrate the circular, which was aimed at Illinois homeowners. By that time Miller had come to identify Simonds's use of predominantly native vegetation at the cemetery as the beginning of a regional "middle western movement" in landscape gardening (known today as the Prairie School)—although Simonds did not share Miller's view.[32] Ultimately, Miller went so far as to call Graceland "perhaps the most famous example of landscape gardening designed by a western man."

In 1925 the Architectural League of New York awarded O. C. Simonds & Company a Medal of Honor in Landscape Architecture.[33] Simonds won the medal for the "excellence in landscape treatment" displayed by photographs of a pair of his projects, one of which was of Lake Hazelmere at Graceland.[34] In 1971, more than half a century after Miller's and Waugh's accolades, the Harvard professor of landscape architecture Norman T. Newton also mentioned Graceland in his now classic reference text, *Design on the Land: The Development of Landscape Architecture.* Like his predecessors, Newton credited Simonds with having single-handedly developed Graceland "into one of the most remarkable park-like cemeteries of the Western world."[35] By the time of Newton's study, however, Graceland's landscape had changed much.

None of these three critics acknowledged the involvement of any landscape gardener other than Simonds in Graceland's design. Waugh, after remarking that "considerable areas of Graceland today [in 1910] present the stone-yard and junk-shop appearance of the usual cemetery," added that these sections were "certainly not part of Mr. Simonds's design."[36] Waugh may not have realized that even these sections were designed by landscape gardeners of repute. William Saunders's original design, for instance, was considered tasteful in its day, although it was unfashionable when Waugh wrote half a century later. When Miller returned his attention to Graceland, this time to promote it as marking the genesis of the Prairie School, he did acknowledge Bryan Lathrop—not as a design contributor, however, but only as Simonds's mentor. Unsurprisingly, Miller made no reference to Jenney or any of Simonds's other landscape gardening predecessors. That Miller, Waugh, and eventually Newton considered Graceland to be Simonds's creation was perhaps the result of their ignorance of the cemetery's long history. Miller's and Waugh's views probably were not informed by careful historical investigation, yet they tended to become entrenched because both were well-known authorities themselves.

Thus the contributions of Nelson, Saunders, Cleveland, Jenney, Cole, and even Lathrop became obscured through the passage of time. By the early twentieth century, in the absence of any scholarly historical accounts of Graceland's development and evolution, and because Simonds's work was recent and fresh and his naturalistic aesthetic in fashion, the older designers had been largely forgotten. In the 1999 textbook *Landscapes in History: Design and Planning in the Eastern and Western Traditions*, by Philip Pregill and Nancy Volkman, the assessment is more even-handed; the authors write: "One of the most important cemeteries of this period [the 1850s and 1860s]— important because of its later influence on the twentieth-century style known as the Prairie Style—was Graceland Cemetery. . . . Although earlier plans have been attributed to William Saunders and H. W. S. Cleveland, it is Ossian Simonds to whom the more famous and still extant scheme is credited."[37]

LANDSCAPE DECLINE

Even as it gained additional monuments, Graceland lost many of its early landscape features. Lotus Pond was the first of substance to go. By 1897 it had been drained and the reclaimed land put to burial use. Such major landscape alterations apparently accelerated after Simonds's death in 1931. Around 1950, for instance, Graceland's railroad station was demolished and the Buena Avenue entrance walled off. Presumably because the cemetery could no longer be accessed from the east, Tanglewood Path was soon replaced with turf, and most of the trees and shrubs that once spatially delineated the corridor were either removed or not replaced after they declined or died. By the 1960s and into the 1970s, Dutch elm disease had begun to deface Graceland's now urban forest, and no tree replacement program was implemented. Lake Hazelmere had been drained by 1968, and this area too was given over to burials; apparently the lake's vital contribution toward Simonds's Medal of Honor, if it was remembered, was not enough to save it.

In the early 1990s the preservation architect John Eifler chronicled the deterioration of the cemetery's landscape in greater detail. Of the period between Simonds's death through the 1980s, he writes:

> Altered cultural perceptions towards death and the deceased, coupled with improved medical practices, overwhelmingly changed public opinion toward cemeteries. Once considered worthy of weekly social visits, garden cemeteries such as Graceland were perceived as dark, foreboding environments to be entered during funerals only. . . . As a result of declining interest, [the landscape] slowly deteriorated for a period of nearly sixty years. Simonds' naturalist environment was eventually forgotten as a succession of superintendents carried on with the business of running the cemetery.[38]

Eifler identified at least five circumstances, all largely beyond Graceland's control, that led to the demise of the cemetery's landscape as Bryan Lathrop and O. C. Simonds knew it. First, atmospheric pollution took its toll on Graceland's vegetation. This, together with public apathy, diminished the need to maintain the landscape to Simonds's standard. Second, security concerns led maintenance crews to prune back tree and shrub foliage heavily in order to remove cover for potential trespassers. Third, the 1950s and 1960s saw a national exodus to the suburbs; Graceland's once-thriving urban surrounds experienced a marked socioeconomic decline. Fourth, relatively fewer descendents of established families were interred at Graceland. Those who did choose to continue family tradition were buried in plots purchased years, if not decades, before. Both the number of new monuments erected and the number of plots sold subsequently declined. Fifth, the rising popularity of cremation also affected revenue, as the number of in-ground burials decreased.

REDISCOVERY AND RENEWAL

By the 1980s Graceland had begun to experience a rediscovery. Eifler notes that a growing interest in American history and culture led many to explore the cemetery, while at the same time it increasingly served as a urban refuge, an oasis of "green space" for those who sought respite from the city's bustle and congestion.[39] Perhaps the earliest signals of Graceland's rediscovery came in the mid-1970s, when the Chicago Architecture Foundation developed walking tours of the cemetery and Graceland made maps available for self-guided tours.[40] Monuments to the famous, rather than the landscape, were the primary attraction for visitors. Nonetheless, interest in Graceland would continue to grow.

In 1991 the trustees of the Graceland Cemetery Improvement Fund commissioned Eifler & Associates, Architects, to formulate a long-range plan for restoring the landscape, buildings, and historically significant monuments. The first landscape outcome was the restoration of a prototype area within the Bellevue section in 1993. Rejuvenation of the Ridgeland section and the O section followed in 1996. (Fig. 8.15) These restoration initiatives continue to the present day. In 2001 Graceland's historical significance was

8.15. The rejuvenated Ridgeland section.
Photograph by William Kildow, Kildow Photography.

8.16. Restored plantings along the edge of Lake Willomere. Photograph by Carol Betsch, 2009.

recognized as national in stature and scope by its inclusion in the National Register of Historic Places.

In 2009, Wolff Landscape Architecture completed a "Strategic Landscape Plan" for the cemetery. Under the firm's direction, the shoreline of Lake Willowmere and the attendant plantings were restored. (Fig. 8.16) As its rejuvenation continues, Graceland's landscape may once again attract visitors as much as its monuments presently do.

Graceland's rediscovery by the public continues. Erik Larson's best-selling book on the World's Columbian Exposition, *The Devil in the White City: Murder, Magic, and Madness at the Fair That Changed America* (2003) prompted many of its readers to visit Graceland. Daniel Burnham and his orchestration of the fair figure prominently in Larsen's narrative, leading many readers to seek out Burnham's monument on the island in Lake Willowmere. More recently, the cemetery inspired the Chicago playwright Ellen Fairey to use it as the setting for her play *Graceland* (2009), which was praised by the *Chicago Tribune* as "haunting" and "beautiful." The cemetery is the venue for a pair of estranged siblings' attempts to come to terms with the mysterious circumstances surrounding their father's death.[41] The drama met with critical acclaim and was later produced at Lincoln Center in New York City.

Today, Graceland Cemetery, as the report prepared by Eifler & Associates noted, has gained enlarged value as "urban green space." (Fig. 8.17) O. C. Simonds actually prophesied this new role. Writing just prior to his death, Simonds observed that burials at the cemetery would "gradually diminish as the families of the present lot-owners disappear." He then asked: "What of Graceland, then— say, one hundred, two hundred, or a thousand years from now?" Simonds hoped it would "continue to serve the living by being a place of quiet retreat, a place of beauty, a place of park-like character. . . [and] venerable trees and exquisite landscapes." Ultimately, he desired the cemetery to be "a place where one who wished to be alone with Nature for a time can satisfy that wish."[42] This is precisely the place Graceland has become. Art historian Sally A. Kitt Chappell recently included the cemetery in her guidebook *Chicago's Urban Nature* (2007). For her, as for many of its admirers, Graceland exemplifies "urban nature" in Chicago, a special place "where architecture and landscape were not only both present but where each had been conceived in response to the other, where the two had been created together as a single artistic whole."[43]

(*Following pages*)
8.17. Aerial view of Graceland in 1992.
Photograph by Robert Cameron, from his *Above Chicago* (San Francisco: Cameron & Company, 1996).

Afterword

In 1991 the chairman of the Graceland Cemetery Trustees' Buildings and Grounds Committee, Robert Isham Jr., embarked on a lengthy inquiry into the archives—including lot cards, correspondence, drawings, blueprints, and photographs—and the resulting comparison of the historical record and the cemetery's present condition made it clear that the historic landscape had been largely lost. In fact, it had been missing for so long that it was likely no one living had seen it in its days of glory. Isham discussed the situation with the architect Robert D. Douglass, and a preliminary professional review of the archival materials ensued. Conducted by the firm Douglass worked for, Eifler & Associates, this initial review led to an in-depth investigation to develop recommendations for restoration and improvement.

The resulting historical report reviewed site plans, landscape plans, and photos; developed a chronology noting significant events and periods; and informed the trustees in detail about Graceland's earlier appearance, its international esteem, and the elements, especially landscape, that were missing or had been seriously compromised. The report also specified monuments, buildings, and sculptures that were deteriorating.

The trustees' review of this history led to important decisions about long-term goals for improvement. These included restoring as far as possible O. C. Simonds's landscape plan as it had existed in the 1920s, repairing monuments, restoring buildings, and improving the efficiency and quality of operations overall. Other goals included developing long-term landscape maintenance programs, identifying and restoring historically important monuments, and determining future building needs and potential uses for unused land. In response, Eifler & Associates made a set of specific recommendations, which was submitted for review in September 1991.

Rather than pursue a master-plan approach to restoration, the trustees chose to hire a landscape architecture firm, Wolff Associates, to first develop a limited "prototype landscape restoration," after which decisions would be made about larger-scale restoration. The plot chosen was four acres in size, surrounding the intersection of Main, Broad, and Woodlawn avenues. The area had historical landscape plans to serve as a guide, and it was representative of the whole in that it included roadways, a mixture of family monuments, raised and flat headstones, and a sample of large, medium, and small plots. In addition, archival planting plans indicated not only layouts but species of plants used. The work of restoring this small landscape included sod removal, development of new planting beds, filling and seeding of beds (around monuments as well), planting of shrubs and perennials to re-create the "outdoor rooms" that Simonds had designed, planting along roads, tree planting in lawn areas and planting beds, and extension and modernization of the irrigation line.

In late 1995, after perennials and groundcovers had filled in and shrubs and trees had attained some significant growth, the trustees, recognizing the skills of cemetery staff, agreed that the new landscaping was affordable, maintainable, and attractive, and that it provided a window on a historical landscape that had not existed since the 1920s. Thus the trustees authorized the restoration of a much larger area, the Ridgeland and "O" sections.

Substantial documentation existed for these sections: extensive planting plans and numerous photos, as well as an overall plan for about 80 percent of the Ridgeland section. But there were impediments to doing a strict restoration. Where Simonds had provided open-area design using sun-loving plants, for instance, that area was now shaded by large, mature trees, so plant species had to be changed to shade-tolerant plants. Other characteristics, such as bed lines and plant masses, were maintained as close to the original as feasible. Landscape installation was completed in the summer of 1996, and today visitors can see again the enclosures and outdoor rooms that provide the quiet and privacy appropriate to the plots, as Simonds had intended.

Lake Willowmere provided the next opportunity for significant landscape restoration, beginning in 2000. There, the major problem was that the intentionally blurred line between the banks and the water surface had been greatly altered by removal of shrubs, by erosion, and by the addition of a rubble retaining wall that created a bright, reflective, and highly visible line between water and land. Over several years the rubble wall was removed, the banks regraded, and plantings added to soften this sharp disjunction. This process went forward section by section, beginning at the southeast portion of the lake and eventually including the entire perimeter. The Burnham Island phase included development of new wetland pockets adjacent to the island, which assisted in the blurring effect and also provided new aquatic habitat.

At the Clark Street entrance the gates had been widened to accommodate vehicles and the entrance to the administration building had been moved from its west to its north facade, but otherwise the administration and waiting room buildings remained beautiful and historically significant. The landscape, however, had greatly changed. Archival photos show ornamental trees flanking the entrance, tall American elms arching above the administration building, shrub masses on both sides of the entry drive narrowing the view into the cemetery, and more shrub masses in the background that hid the headstones and monuments from immediate view and created a sense of mystery. With a few variations for the sake of security and safety—such as holding back plantings from the edges of the roadway instead of allowing them to spill over—all of these features were restored. The new elms are disease-resistant hybrids.

In addition to these focused landscape restorations, tree planting throughout the cemetery has been extensive, especially on the perimeter, to increase the sense of enclosure and separation from surrounding neighborhoods, where buildings have tended to increase in height over the years.

Building construction and restoration began in 1995 when the trustees recognized that the basement of the Holabird & Roche–designed chapel, where bodies had been stored when the ground was frozen, was functionally obsolete, and so it was filled. However, new space was needed for cremated remains, so the decision was made to build a new columbarium near the chapel. Working together, the Eifler and Wolff firms designed a small pool, beyond which are irregular garden rooms formed by curving granite walls that replicate the color and texture of the Waupaca granite of the adjoining chapel. Urns may be interred either in front of or behind the walls, or cremated remains may be scattered; in either case a memorial plaque is affixed to the wall.

In 2008 the trustees authorized the restoration of the chapel, beginning with the demolition of the two additions on the north side and continuing with the rehabilitation of the chapel itself. One of the adaptations required by the chapel restoration was a new route from the drive to the entrance on the north side of the building. The demolition of the additions also opened up a substantial new area for interments, and the trustees chose a development similar to the Ridgeland section recently completed: small family plots suitable for modestly sized individual monuments, with spatial enclosure and separation provided by plantings at the backs and sides of the plots, in the style of Simonds but at smaller scale. This design, which focuses on an open central lawn area extending north from the chapel, was completed in 2010.

Ted Wolff
Wolff Landscape Architecture, Inc.

Photographs on pages 194–204 are reproduced from *Graceland Cemetery* (Chicago: Photographic Print Co., 1904), courtesy Trustees of the Graceland Cemetery Improvement Fund. Captions are original.

The Chapel Grounds.

Opposite: **The Chapel Elms.**

Main Avenue.

Maple Avenue.

Lake Avenue.

Woodlawn Avenue.

Ridgeland Section.

Edgewood Section.

Opposite: **Maplewood Section.**

Fairlawn Section.

The Knolls.

Willowmere.

Single Graves.

Greenwood Avenue.

Willowmere.

Willowmere.

Edgewood Section.

Hazelmere.

Acknowledgments

Nearly thirty years ago, while pursuing a degree in landscape architecture, I first gained exposure to Graceland Cemetery in a history class. This introduction came through poorly reproduced and dimly projected black-and-white slides of photographs taken in the late nineteenth and early twentieth centuries. Many of these views, I later discovered, were the work of an important landscape photographer, Arthur G. Eldredge, and were made to illustrate critic Wilhelm Miller's publications, including his celebrated design manifesto *The Prairie Spirit in Landscape Gardening* (1915). Despite the graininess of the reproductions and perhaps in testament to Eldredge's talents, the images nevertheless compelled me to visit the cemetery. Graceland's reality in the 1980s, however, proved a profound disappointment; the cemetery's lushly planted landscape had markedly declined since the time Eldredge documented it. Instead of trees and shrubs, monuments and other markers now predominated. But this diminished state became a somewhat poignant catalyst, prompting me to investigate Graceland's beginnings and the evolution of its landscape design. And, ever inspired and encouraged by my late mentor Walter L. Creese (1919–2002), I soon launched what would prove to be a years-long effort to recover Graceland's past. Although I did not foresee it, nearly a decade (and a master's degree) later I would contribute professionally to efforts to rejuvenate Graceland's botanical mantle. No less unanticipated, nearly a decade after that, an invitation to chronicle the cemetery's design history in print found me "down under" in distant Australia (a move made initially to follow the trail of Walter and Marion Griffin, the Chicagoan designers of Australia's national capital, Canberra; coincidentally, Marion's cremated remains are interred at Graceland).

Along with multiple marathon flights to Chicago to consult archives and study Graceland's now once again flourishing landscape firsthand, conducting research across two hemispheres inevitably requires collegial assistance and collaboration. First and foremost, I gratefully acknowledge the tireless assistance of John K. Notz, Jr. A member of the Trustees of the Graceland Cemetery Improvement Fund and a driving force behind this project, John first contacted me in Australia at the suggestion of Frank Lloyd Wright scholar Anthony Alofsin, seconded by Leonard K. Eaton, one of Walter Creese's dear friends and the pioneering biographer of the now-legendary Prairie School landscape architect Jens Jensen. As is often the case with books, this one took longer than anticipated to write, so I must also thank John for his patience. Thanks are also due him and his wife, Janis, for their generous hospitality during my Chicago sojourns. It was a pleasure to again work with Robin Karson; I am grateful to her not only for her editorial assistance, but also for her ongoing support and timely encouragement. Also invaluable were the skills of developmental editor Joel Ray (who also kindly provided me with archival assistance at Cornell University), project editor Mary Bellino, and illustration researcher Jessica Dawson. Carol Betsch's poetic sense of place is evident in new photographs that appear throughout the book, bringing the Graceland story up to the present day. At Graceland Cemetery, my work was materially aided by Aki Lew and Diane Lanigan. The Chicago History Museum, which holds many of Graceland's historical records, also provided expert assistance; I offer special thanks to Lesley Martin and Debbie Vaughan of the museum's Research Center.

I also thank the many librarians, archivists, and fellow scholars in Chicago and elsewhere in the United States who kindly answered my requests, no doubt puzzled to receive e-mail messages from Australia inquiring about the history of a cemetery in Chicago. In the Windy City, Robert Karrow at the Newberry Library, Julie Lynch at Sulzer Regional Library, Julia Bachrach at the Chicago Park District, Michael Conzen at the University of Chicago, and architect David Swan all helped me further my research. The Internet, of course, greatly facilitated, if not enabled, my working at such a great distance. Art historian Wendy Greenhouse shared her critical insights, and along with other forms of assistance she secured my access to the historical *Chicago Tribune*. This online resource, unavailable to earlier writers on Graceland, yielded much important new documentation on the cemetery, virtually irretrievable by any other means. Nancy Wilson at the Elmhurst Historical Museum helped me track down clues to the landscape gardening activities of Graceland's founder, Thomas Barbour Bryan. John Reinhardt of the Illinois State Archives and Jane Ehrenhart and Cheryl Schnirring of the Abraham Lincoln Presidential Library, both in Springfield, were also of great assistance. At Illinois State University, Jo Ann Rayfield and Bruce Stoffell of the Milner Library fulfilled my requests for information on early Graceland landscape gardener William Saunders's plan for that university's campus. Kathleen Seusy, Neil Dahlstrom, Gretchen Small, and Todd Slater answered my many queries concerning the John Deere family and William Le Baron Jenney's work in Moline, Illinois. At the University of Michigan, Marilyn McNitt and Jennifer Sharp of the Bentley Historical Library aided my efforts to investigate William Le Baron Jenney's and O. C. Simonds's time in Ann Arbor. Emily Romick, Joel Fry, Laura Beardsley, Emily T. Cooperman, and John Dixon Hunt assisted me in retracing the paths of William Saunders and John Jay Smith, founder of Laurel Hill Cemetery in Philadelphia. Mike Klein, at the Geography and Map Division of the Library of Congress, located historical maps that depict Graceland. Also in Washington, Nancy Witherell of the National Capital Planning Commission enhanced my appreciation of Thomas Bryan's later role as one of the commissioners of the District of Columbia. Elsewhere, colleagues William H. Tishler, Robert E. Grese, and Aaron Wunsch not only assisted, but also shared their insights into the project. At the University of Western Australia, I am grateful to Dr. Clarissa Ball, formerly dean of the faculty of Architecture, Landscape, and Visual Arts, for her support and encouragement. As this book has been years in the making and has received generous support from so many individuals, any omissions are not intentional; they are purely the outcome of a faulty memory.

I especially thank my partner, Annette Condello, who was and remains my foundation. Ethereally, long before I began this project and before we became a couple, she had an enigmatic dream in which we were strolling together through an unknown cemetery, a hemisphere away in America. Little did either of us know that we would realize her dream at Graceland and that it would prove a harbinger of this book.

Lastly, this book is dedicated to the memory of Walter L. Creese. In 1920, O. C. Simonds similarly dedicated his own book to the memory of his mentor Bryan Lathrop, formerly president of the Graceland Cemetery Company. Lathrop's influence, Simonds wrote, had been "felt in each page of this volume and in all the professional work of the author." Walter's influence on this writer's work is no less, and he remains sorely missed.

Christopher Vernon
University of Western Australia

Notes

Introduction

1. Miller, *Prairie Spirit in Landscape Gardening*, 2.
2. Ibid.
3. Bruegmann, *Architects and the City*, 3. The name *Chicago* is thought to be a French variant of a Native American term for "wild onion."
4. Eslinger, "Gardening."
5. D. Miller, *City of the Century*, 88.
6. Bruegmann, *Architects and the City*, 3.
7. Karamanski, "Civil War"; see also his *Rally 'Round the Flag*.
8. Bruegmann, *Architects and the City*, 4.
9. Sclair, "Cemeteries."
10. Chappell, *Chicago's Urban Nature*, 154.
11. Ibid.
12. D. Miller, *City of the Century*, 284.
13. Fuller's *With the Procession* (1895), quoted in Andrews, *Architecture, Ambition, and Americans*, 199.
14. Chappell, *Chicago's Urban Nature*, 154.
15. Ibid.
16. Sullivan, *Autobiography of an Idea*, 243–44.
17. Hubbard and Kimball, *Introduction to the Study of Landscape Design*, 64–65.

1. Thomas Barbour Bryan and the Genesis of Graceland

1. *National Cyclopaedia of American Biography*, 3:170. See also the entries in Johnson and Malone, *Dictionary of American Biography*, 2:190–91; and *Who Was Who in America*, 1:153. Chapter 8 in Berens, *Elmhurst* (77–95), is devoted to Bryan. Berens writes: "As an undergraduate he specialized in languages: German, French, Italian, Latin and Greek. Before his graduation he wrote a novel text-book 'For Germans more easily to learn English,' which was published by Appleton's and ran through several editions" (79). In Chicago, the Bryans' addresses included Michigan Avenue at Madison Street, Wabash Avenue at Jackson Street, and Division Street at Lake Shore Drive (79). Bryan's father, Daniel Bryan, also an attorney and a state senator in Virginia, was a poet whose verses won the respect of Edgar Allan Poe; see Binns, "Daniel Bryan."

2. Johnson and Malone, *Dictionary of American Biography*, 2:190–91.

3. "Gallery of Local Celebrities, No. XVI. – Thomas B. Bryan," *Chicago Tribune*, 13 May 1900, 39.

4. Ibid.; see also Russell, *Elmhurst*, 25.

5. Funigiello, *Florence Lathrop Page*, 16; and "Gallery of Local Celebrities."

6. The hall was at 119–21 North Clark Street. For the site's later history see "RKO Grand Theater," in Randall, *Building Construction in Chicago*, 71.

7. *Bryan Hall, Clark Street, Opposite the Court House, Chicago, Illinois* (1860), Chicago History Museum, "Programs, unbound and arranged by years, 1861–1862"; it may also be seen at http://hdl.loc.gov/loc.rbc/rbpe.01807500. In this promotional broadside, "Real Estate Attorney and Counsellor at Law" Bryan describes the building in detail and also advertises his real estate investment practice.

8. "Bryan's New Music Hall—Noble and National Art Decorations," *Chicago Press and Tribune*, 20 March 1860, 1. See also the following articles in the *Chicago Press and Tribune:* "Bryan Hall," 5 September 1860, 1; "The Dedication of Bryan Hall," 18 September 1860, 1; "Bryan Hall and Portrait Gallery," 18 October 1860, 1. On the building's frescoes, by the well-known Chicago firm of Jevne & Almini, see "House Painting," 27 July 1860, 1.

9. Berens, *Elmhurst*, 80. Berens writes that Bryan Hall "was the chief center of musical and theatrical activities, as well as a hub of the social life of the city. Here Chicago society held its great balls and masquerades; here appeared the leading singers, musicians, and entertainers of the time; and here the leading singing societies gave their concerts."

10. "Ralph Waldo Emerson's Lecture," *Chicago Tribune*, 23 January 1863, 4.

11. Berens, *Elmhurst*, 80.

12. Laura Kendall Thomas, "Story of Hill Cottage, Later Called Cottage Hill Tavern," typescript dated 18 November 1936, Elmhurst Historical Museum, Elmhurst, Ill. I thank Nancy Wilson, archivist at the Elmhurst Historical Museum, for supplying me with a copy of this document. According to Thomas, Bryan made the purchase in October 1856. See also Russell, *Elmhurst*, 13–35; and "Cottage Hill," *Chicago Tribune*, 4 October 1867, 2.

13. Berens identifies the site as the Peter Fippinger farmhouse, on County Line Road, where the Bryans lived for a short time in 1857 and 1858. Berens, *Elmhurst*, 81. In 1936 Elmhurst pioneer Wilbur Hagans recalled: "At the time of my arrival in Cottage Hill, Nov. 9, 1857, Thomas Bryan and family were living at the extreme western edge of Cook County, in a one story frame house at what was then known as 'The Grove.'" See Bio 1 – Bryan, Thomas (biographical file), Elmhurst Historical Museum, hereafter cited as Bryan file, EHS. See also "Fippinger Family" and "Life in 'Proviso' (1850–1900)" at www.franzosenbuschheritageproject.org. This site also includes a digital reproduction of a plat map titled "Land Ownership in 1863." Bryan's property is identified in Section 7 of Proviso Township, Cook County. I am grateful to historians Nancy Wilson at the Elmhurst Historical Museum and Shirley Slanker of the Franzosenbusch Heritage Project for their assistance in documenting Bryan's land holdings.

14. Wilbur Hagans dated the beginning of the construction of a brick house on the Cottage Hill property to the spring of 1858, with the Bryans moving in the following year. Bryan file, EHS.

15. Thomas Barbour Bryan, letter to "My Dr Mr Wylie," 26 June 1858, Thomas Barbour Bryan letters [manuscript], 1853–1889, MSS Alpha1 B, Chicago History Museum. Bryan's brother-in-law Andrew Wylie (1814–1905) was a

lawyer who later became a judge of the Supreme Court of the District of Columbia.

16. Russell, *Elmhurst*, 25–26; Russell notes that the completed house "may be said to mark the beginning of Elmhurst as a suburb."

17. "How Chicago Suburbs Were Planted and Named," *Chicago Tribune*, 25 February 1900, 33.

18. Berens describes the fountain, which was purchased in Florence: "This marble fountain had a pedestal made up of cupids holding the large shell which formed a bird-bath on top of the fountain. It was so constructed that it connected with the water system on the property. Occasionally when there were visitors to the Bryan gardens, and often on Sunday afternoons, the fountain would be turned on. A large stream of water would shoot up in the center of the basin, cascading down, to the delight of the visitors" (*Elmhurst*, 110).

19. Wilbur Hagans recalled that the landscape architect at Bird's Nest was Swain Nelson. Bryan file, EHS. More likely, however, Nelson was hired to implement Bryan's own design.

20. "Rose Hill Cemetery," *Chicago Press and Tribune*, 9 March 1859. On John J. Smith, see Aaron Wunsch's entry on him in Birnbaum and Karson, *Pioneers of American Landscape Design*, 373–75. Smith served as librarian for the Library Company of Philadelphia from 1829 to 1851.

21. On Downing, Smith, and the *Horticulturist* see Judith K. Major's excellent monograph, *To Live in the New World*.

22. B[ryan], "Defence of the Prairies."

23. Ibid.

24. S[mith], "Our Chicago Correspondence," 244.

25. Olmsted, "Preliminary Report upon the Proposed Suburban Village at Riverside," 276.

26. S[mith], "Our Chicago Correspondence," 245.

27. B[ryan], "Defence of the Prairies."

28. "The Farm and Garden: Cottage Hill," *Chicago Press and Tribune*, 6 July 1859, 2.

29. Thomas, "Story of Hill Cottage." Thomas cites local property records showing that Healy purchased "Blocks 15 and 16 Summit Addition to Cottage Hill for $8,510.00" from Bryan in August 1857; the frontage is described as being "on St. Charles Road from 'The Avenue' (now Cottage Hill Avenue) to York Street. The old tavern [the namesake cottage] stood on St. Charles at the west end of this property." She reports that Healy's youngest daughter, Mrs. Charles H. Besley, later recalled that "it was Mr. Bryan who induced my father to buy out there" (7–8). See also Russell, *Elmhurst*, 16. Healy also reported that his children's health was a motivation for his move to Cottage Hill; see Healy, *Reminiscences*, 67. Healy lived at Cottage Hill from 1857 until 1863; that year he returned to Boston. He next departed for Rome and then maintained a Paris studio from 1885 to 1892, when he returned to Chicago; he died there in 1894. Russell, *Elmhurst*, 19.

30. Berens, *Elmhurst*, 37–38. Again citing Healy's youngest daughter, Laura Kendall Thomas writes: "Stretching to York Street was the lawn of pink clover which gave the place the name, 'Clover Lawn'" ("Story of Hill Cottage," 9).

31. Bryan Lathrop, letter to John F. Cremin, 22 September 1897, Elmhurst Historical Museum. I thank archivist Nancy Wilson for supplying me with a copy of this letter. Lathrop notes that Huntington was laid out between 1864 and 1866 by Swain Nelson, including the transplanting of many large

elms. For further details see Russell, *Elmhurst*, 28; and Berens, *Elmhurst*, 40.

32. "Gallery of Local Celebrities." On tree planting in Elmhurst see, for example, Bates, *"Old Elmhurst,"* 18.

33. For more on Elmhurst see Knoblauch, *Du Page County*; and Elmhurst Historical Commission, *Elmhurst: Scenes from Yesterday*.

34. On Smith and Saunders's presence in Chicago see "Rose Hill Cemetery," *Chicago Press and Tribune*, 9 March 1859, 1. (The name is variously spelled Rosehill and Rose Hill in early sources, but in the cemetery's charter it is given as Rosehill.)

35. [Smith], "Editor's Table" (April 1859), 185.

36. S[mith], "Our Chicago Correspondence," 245.

37. For a comprehensive account of the origins and evolution of rural cemeteries see Linden, *Silent City on a Hill*.

38. Eggener, "Building on Burial Ground," 9.

39. Wunsch, "Emporia of Eternity," 16.

40. Ibid.

41. See Andreas, *History of Chicago*, 2:449.

42. Andreas, *History of Chicago*, 1:141–42 and 2:448. Bryan subdivided the tract into residential lots and specified that each house be set back a hundred feet from the street.

43. On Rauch's medical activities see Davenport, "John Henry Rauch."

44. Rauch, *Intramural Interments*, 23. Although *Intramural Interments* was published in 1866, in his preface Rauch noted that the text "was prepared in the fall of 1859."

45. Ibid., 62; see also Rauch, *Public Parks*.

46. Rauch, *Intramural Interments*, 63.

47. On the earlier efforts toward relocation see "A New City Cemetery," *Chicago Tribune*, 18 February 1857, 1.

48. "The Day at Rosehill: Inauguration of the New Cemetery," *Chicago Press and Tribune*, 29 July 1859, 1.

49. *Charter, Rules and Regulations of the Rosehill Cemetery*, 15, 30. See also "Report on the Cemetery Matter," *Chicago Press and Tribune*, 16 February 1859, 1.

50. "Rose Hill Cemetery," *Chicago Press and Tribune*, 9 March 1859, 1.

51. See "A Great Desideratum Supplied—Rosehill Cemetery," *Chicago Press and Tribune*, 27 July 1859, 1; and "The Day at Rosehill," *Chicago Press and Tribune*, 29 July 1859, 1.

52. See "The New Catholic Cemetery on the Lake Shore," *Chicago Press and Tribune*, 23 August 1859, 1; and "Opening of Calvary Cemetery," *Chicago Press and Tribune*, 4 November 1859, 1.

53. "Died," *Chicago Tribune*, 13 April 1855, 2.

54. Thomas Bryan, "Our Two Cemeteries: 'Now, What Says Graceland?'" (letter to the editor), *Chicago Tribune*, 8 December 1862.

55. Ibid.

56. Bryan reported the purchase in a leaflet he later prepared for Graceland's dedication ceremony; see Graceland Cemetery, *Formal Dedication*, broadside collection, Chicago History Museum. The purchase is similarly documented in "Organization & Election of Officers, Graceland Cemetery Co." (classified advertisement), *Chicago Press and Tribune*, 27 June 1860, 1. "Graceland Cemetery," another advertisement, includes a more detailed description of the site, quoted from Saunders's initial evaluation; *Chicago Press and Tribune*, 6 April 1860, 1. It is unclear whether Bryan purchased the property directly from Justin Butterfield (1790–1855) himself or his heirs. A pioneering Chicago lawyer, land speculator, and commissioner of the General Land

Office under President Zachary Taylor, Butterfield died in October 1855, the same year Bryan's son died. It is possible that at some point between his son's death in April and October 1855, Bryan purchased the land from Butterfield. If he did, then perhaps it was an unfavorable economic climate that conspired against proceeding with the cemetery project until now. On Butterfield see Andreas, *History of Chicago*, 1:433–35.

57. Graceland Cemetery, *Formal Dedication*.
58. [Graceland Cemetery Company], *Charter of the Graceland Cemetery*, n.p.
59. Kirkland, *History of Chicago*, 2:603.
60. Hudson, "Topography."
61. See Pattison, "Land for the Dead," 49–52, and his map titled "Natural Setting of Chicago Cemeteries" (fig. 2).
62. [Graceland Cemetery Company], *Catalogue of the Graceland Cemetery Lot Owners*, 5. The report is dated 2 August 1860 and is signed by surveyors Samuel S. Greeley and Edmund Bixby.
63. [Graceland Cemetery Company], *Charter of the Graceland Cemetery*, n.p.; see also "The Dedication of Graceland Cemetery," *Chicago Press and Tribune*, 31 August 1860, 1, and "Our Burial Places," *Chicago Tribune*, 11 December 1862, 4.
64. Reynolds, *Limit of the Police Power*, 40.
65. See Abrahamson, "Lake View."
66. Seligman, "Uptown."

2. The First Designers: Swain Nelson and William Saunders

1. Thomas Bryan, "Our Two Cemeteries: 'Now, What Says Graceland?'" (letter to the editor), *Chicago Tribune*, 8 December 1862.
2. Simo, *Loudon and the Landscape*, 282–83.
3. Wunsch, "Emporia of Eternity," 21, 23n25. The extended subtitle of Smith's volume suggests as much; it reads *With a Preliminary Essay on the Laying Out, Planting and Managing of Cemeteries and on the Improvement of Church Yards, on the basis of Loudon's Work*.
4. "New Publications: Practical Landscape Gardening," *Chicago Tribune*, 2 April 1855, 3.
5. Kern, *Practical Landscape Gardening*, 263.
6. Ibid., 265–66.
7. On Loudon and the contrast between the gardenesque and the picturesque see Simo, *Loudon and the Landscape*, 165–90.
8. It is unclear whether G. M. Kern is the same person as the German-born landscape gardener Maximilian G. Kern (late 1820s/early 1830s–1915), who may have spent time in Cincinnati in the mid-1850s. See Richard Longstreth's entry on Maximilian Kern in Birnbaum and Karson, *Pioneers of American Landscape Design*, 209–12.
9. Kern, *Practical Landscape Gardening*, 264. Kern mentions these cemeteries not for their designs but as "specimens of finely chosen situations."
10. The article is attributed only to "A. D. G., Clinton, N.Y." Edward North identified the author as the "Rev. A. D. Gridley" in his own article, "The Proper Expression of a Rural Cemetery," in the June 1857 issue. Another article on rural cemeteries published in the *Horticulturalist* around this time is W. Scott, "Village Cemeteries," which appeared in April 1855.
11. G[ridley], "Rural Cemeteries," 280.
12. Nelson came to America in 1854; prior to moving to Chicago he worked in Defiance, Ohio. For more on him see Maloney, *Chicago Gardens*, 12, 17, 356; and "Autobiography of Swain Nelson (1828–1917)," www.gyllenhaal.org/

SwainNelsonAutobio.html. A copy of the autobiography is also available in the Academy of the New Church Archives, Bryn Athyn, Pa.

13. "Autobiography of Swain Nelson." The autobiography was based on Nelson's recollections as dictated to his granddaughter.

14. "J. B. Waller" (obituary), *Chicago Tribune*, 5 August 1887, 2. The obituary notes that Waller "removed in 1858 to the vicinity of Chicago"; Nelson mentions the size of Waller's parcel in his autobiography. The house Waller built, named Buena Vista, was described in 1874 as looking "like the fine old mansions to be seen in long-settled districts of the East." "The cupola commands a view of Lake Michigan and a large radius of country around. The material of the house is brick. The interior is elaborately finished in hardwood, and its fine apartments are spacious, pleasant and comfortable. The cost of the structure was about $75,000." Chamberlin, *Chicago and Its Suburbs*, 351.

15. "The Dedication of Graceland Cemetery, *Chicago Press and Tribune*, 31 August 1860, 1.

16. [Graceland Cemetery Company], *Charter of the Graceland Cemetery*, n.p. An 1892 biographical account of Bryan also confirms him as not only Graceland's "originator," but also its "sole proprietor (formerly)"; *Biographical Dictionary and Portrait Gallery*, 39.

17. "The Dedication of Graceland Cemetery," *Chicago Press and Tribune*, 31 August 1860. See also Ware, "Memorial of William Saunders"; and the entry for Saunders in *Who Was Who in America: Historical Volume*, 465. At the time of the Graceland commission, Saunders had just completed his competitive design for Fairmount Park; his published report, *Design for Fairmount Park*, is dated 12 February 1859. Another source notes that in 1859 Saunders "was engaged in laying out the Rosehill Cemetery in Chicago, and another at Evanston, Ill." The mention of a cemetery at Evanston is likely a reference to the Roman Catholic cemetery Calvary; however, it might possibly allude to Graceland. See the entry for Saunders in *Yearbook of the United States Department of Agriculture, 1900*, 625–30.

18. *Yearbook of the United States Department of Agriculture*, 625.

19. See Reuben Rainey's entry on Saunders in Birnbaum and Karson, *Pioneers of American Landscape Design*, 327.

20. Brackett, "Horticulturists: Saunders, William." A year later, however, the partnership was dissolved. See *The Florist and Horticultural Journal*, August 1855, 256.

21. *Yearbook of the United States Department of Agriculture*, 626.

22. In October 1859 John J. Smith noted the Rahway commission in the editor's column of the *Horticulturist*, describing the fifty-acre site and predicting that "as in all previous engagements in landscape gardening, Mr. Saunders will give entire satisfaction." [Smith], "Editor's Table: Rural Cemeteries and Public Parks," 568. Saunders's promotional leaflet "Landscape Gardening and Rural Improvements. Wm. Saunders, Landscape Gardener" included a testimonial from the "Hazlewood [*sic*] Cemetery Company" that Saunders's design for them combined "the beauties of landscape gardening with perfect adaptability to the purposes of a cemetery." Smith Papers, John Jay Smith Correspondence, Library Company of Philadelphia; the leaflet's original location, before the papers were unbound, was vol. 14, p. 242.

23. For instance, the *New York Times* erroneously reported in its obituary that Saunders was "famed as a landscape gardener, having planned Fairmount Park in Philadelphia." "William Saunders," *New York Times*, 12 September 1900, 7.

24. Saunders listed Ridgely and Kennicott as his clients in his promotional leaf-let (see note 22); Kennicott's residence was named Kenwood. As noted pre-viously, mention of an unidentified cemetery at Evanston probably alludes to Calvary cemetery, but it might also possibly be a mistaken reference to Graceland, as that cemetery is similarly located to the city's north. See *Year-book of the United States Department of Agriculture*, 626. On Saunders and Oak Ridge see [Walker], *Oak Ridge Cemetery*, 19.

25. Marshall, *Grandest of Enterprises*, 48, 62, 114. Saunders's authorship of the campus is also documented in *Proceedings of the Board of Education of the State of Illinois . . . December 16th, 1868*, 11, and *Proceedings of the Board of Education of the State of Illinois . . . June 6, 1917*, 28. In a letter to Fell dated 15 October 1858, Saunders reported the campus "plan is finished"; Papers of Jesse W. Fell, 1806–1957, Library of Congress, microfilm, Ms 77-1518, reel 2, frame 409. He submitted his "plan for Normal School grounds" to Fell with a covering letter dated 29 October 1858 (reel 2, frame 411). The Fell papers also include a plan view lithograph of Saunders' "Design for the Grounds of the State Normal School, Bloomington, Illinois" (reel 5, frames 473–75).

26. [Graceland Cemetery Company], *Charter of the Graceland Cemetery*, n.p.

27. Wunsch, "Emporia of Eternity," 15.

28. Ibid.

29. See, for instance, Saunders's "Designs for Improving Country Residences," and "Plan of Hunting Park."

30. The precise date Bryan commissioned Saunders is unclear, but it is impor-tant to recognize that contact between the two was fairly extensive before Saunders made his survey of the Graceland site in April 1860, so I will summarize what is known. Records on the Chicago end are scanty, but Saunders kept meticulous correspondence registers and account books, which are included among the William Saunders Papers (cited as WSP) within the National Grange of the Patrons of Husbandry Records, Cornell University Library, Ithaca, N.Y. As we have seen, Bryan had made personal contact with Saunders by June 1858, when Saunders and Smith visited Chi-cago. Two months later, Saunders exchanged two sets of letters with Bryan, apparently among the earliest written communications between the two; the letters themselves are lost, but it is possible that Bryan's rural cemetery project was the subject. (Correspondence Register, box 4, folder 4, WSP; the letters were sent on 2 August and 16 August and Bryant's replies were received on 9 August and 20 August.) By March 1859 Saunders had taken up the Rosehill commission, and in June, according to his account register, that project took him to Chicago for a six-day visit, certainly long enough for Bryan to meet with him again. (Account Register, #3020, box 8, folder 1, WSP, contains billing records for the Rosehill work at this time.) One of Bryan's letters indeed suggests that this was the case: the next month, in a letter to his brother-in-law written on July 20, Bryan somewhat cryptically alludes to his own association with a "Cemetery rumour." On 17 November 1859 Bryan reported in a letter to his parents that he had visited Philadel-phia that month, and although he did not record the reason, perhaps it was to consult Saunders and inspect Laurel Hill. Only a few weeks later, whether a consequence or simply a coincidence, Bryan awarded Saunders the commission: in another letter to his brother-in-law, this one written on 23 December 1859, Bryan reports that there is "nothing new" to say about his "cemetery plan," as "Saunders is to determine the whole matter." These three letters are preserved in the Thomas Barbour Bryan letters [manu-script], 1853–1889, MSS Alpha1 B, Chicago History Museum, hereafter

cited as Bryan Letters, CHM. That Saunders had the commission in hand by January 1860 is confirmed by his record of receiving a check from Bryan. When he deposited it on January 14, however, he noted in his ledger the check was "to be repaid, as [Bryan] ha[d] not yet received value for it" (Account Register, #3020, box 8, folder 1, WSP). This somewhat ambiguous annotation likely means Bryan cancelled payment as Saunders had not yet begun the work.

31. WSP, box 2, folder 2.

32. Healy's purchase was made in compliance with Bryan's plan to clear the artist's considerable debts. Healy detailed the scheme in a January 1860 letter to a friend: "Mr. Thomas B. Bryan offered me 80 acres 4 miles north of Chicago river where it runs through the city half a mile from the Lake at $400 per acre. He proposed to take my Cottage Hill property at $10,000; thirty [portraits] . . . for $10,000. This leaves me $12,000 to pay in cash, to pay which he gives me three years, and during those three years I am to occupy Cottage Hill free of rent. . . . The papers were signed on the 28th." This letter is quoted in his granddaughter Marie De Mare's monograph *G. P. A. Healy, American Artist*, 187. See also Bryan to "Mr. Wylie," 17 February 1860, Bryan Letters, CHM.

33. "Graceland Cemetery" (classified advertisement), *Chicago Press and Tribune*, 6 April 1860.

34. Bryan, letter to John Withers, 12 May 1860, Bryan Letters, CHM.

35. "Organization & Election of Officers. Graceland Cemetery Co." (advertisement), *Chicago Press and Tribune*, 27 June 1860, 1.

36. Healy is listed as treasurer in the 1861 *Charter of the Graceland Cemetery*, n.p.

37. Jack Harpster, 9 February 2008, e-mail to author.

38. See Healy, *Reminiscences*, 57. See also De Mare, *G. P. A. Healy*, 175–84, and Bigot, *Life of George P. A. Healy*, 29–32.

39. Gerdts, "Chicago Is Rushing Past Everything," 47.

40. "Death of Dr. Sidney Sawyer," *Chicago Tribune*, 13 July 1894, 7.

41. See the entry on Butterfield in Andreas, *History of Chicago*, 1:434.

42. Ebert, "Early History of the Drug Trade of Chicago," 271–72. Ebert gives the date of Sawyer's arrrival in Chicago as 1839.

43. Healy, *Reminiscences*, 59–60.

44. On Healy's historical paintings see Voss, "Webster Replying to Hayne."

45. I am indebted to art historian Dr. Wendy Greenhouse for this insight.

46. On the Boston studio see Healy, *Reminiscences*, 27.

47. Untitled advertisement, *Chicago Press and Tribune*, 27 June 1860, 1.

3. The Earliest Designs

1. Saunders, "Plan of Hunting Park," 462.

2. Saunders, "Design for Fairmount Park," 12 February 1859, 2. A copy of this printed and bound report is available at the Boston Public Library. See also Lewis, "The First Design for Fairmount Park."

3. Graceland Cemetery Collection (1994.146), Series I: Maps and Plats, map folder 1, Chicago History Museum. Handwritten annotations indicate that this particular copy of the lithograph was used to record burials; thus it was kept on site at the cemetery rather than at the company's city office, and for that reason it was one of the very few pre-1871 Graceland records to have escaped the Great Fire.

4. Charles Shober (b. ca. 1831), who moved from Philadelphia to Chicago in 1858, became one of the city's most eminent mapmakers. He remained active there until around 1883. See Michael P. Conzen's invaluable survey,

"Evolution of the Chicago Map Trade"; see also Groce and Wallace, *The New-York Historical Society's Dictionary of Artists in America*, 577; and Ristow, *American Maps and Mapmakers*, 417–20 and 432–33. A contextual appreciation of Shober's work can be gained from John W. Reps's monumental study, *Views and Viewmakers of Urban America.*

5. On grottos see, for example, the entry in Taylor, *Oxford Companion to the Garden*, 202–3.

6. Naomi Miller documents a curious exception. In 1871, as part of a proposal for the yet-to-be-built Metropolitan Museum of Art in New York, the painter William H. Beard made an unexecuted plan for a "Grotto Entry" that would link the new museum to Central Park by means of an underground passage. See Miller, *Heavenly Caves*, 120; for further details and illustrations see "An American Museum of Art: The Designs Submitted by William H. Beard," *Scribner's Monthly*, August 1871, 409–15.

7. "Graceland Cemetery," Graceland Cemetery Collection (1994.146), Series I: Maps and Plats, map folder 2 (Chicago History Museum).

8. "Our Burial Places: Graceland, Rosehill and the City Cemeteries," *Chicago Tribune*, 11 December 1862, 4. The same article noted that at Rosehill, the excavation of "a lake of more than two acres in extent" was under way, to be completed the following year, 1863. Although Rosehill was chartered earlier, the two cemeteries developed more or less contemporaneously.

9. Ibid.

10. See, for example, Saunders's advertisement "Rural Improvements," *Horticulturist* 15, no. 1 (January 1860): 14.

11. "Our Burial Places. Graceland, Rosehill and the City Cemeteries."

12. Thomas Bryan, "Our Two Cemeteries: 'Now, What Says Graceland?'" (letter to the editor), *Chicago Tribune*, 8 December 1862, 4.

13. Bryan to John Withers, 12 May 1860, Chicago History Museum, "Thomas Barbour Bryan letters [manuscript]," 1853–1889, MSS Alpha1 B.

14. Ibid.

15. "Organization & Election of Officers. Graceland Cemetery Co." (classified advertisement), *Chicago Press and Tribune*, 27 June 1860, 1.

16. "The Dedication of Graceland Cemetery," *Chicago Press and Tribune*, 31 August 1860, 1.

17. "Our Burial Places, " *Chicago Tribune*, 11 December 1862, 4.

18. "The Cemeteries of Chicago," *Chicago Tribune*, 27 June 1869, 3.

19. "Our Burial Places," *Chicago Tribune*, 11 December 1862, 4.

20. On Carter and Bauer, see Geraniotis, "German Architectural Theory and Practice"; see also "Asher Carter" (obituary) in *American Architect and Building News;* and Wight, "A Portrait Gallery."

21. "Progress of Chicago," *Chicago Tribune*, 4 January 1856, 1.

22. See "New Vault at Graceland Cemetery" (classified advertisement), *Chicago Tribune*, 24 September 1866, 1.

23. Reproductions of some of the architectural drawings for an "Addition for Mr. J. H. Lathrop Residence near Elmhurst Station" survive; they are signed "Bauer & Loebnitz Archts," as Bauer was by then in partnership with another German architect, Robert Loebnitz. Although the drawings are undated, we know that Cottage Hill was renamed Elmhurst in 1869 and that the partnership spanned the years 1866 to 1874. Burnham Library–University of Illinois Project to Microfilm Architectural Documentation Records, 1950–1952 (accession no. 1952.1), reel 37, frames 105–17, Ryerson and Burnham Archives, Ryerson and Burnham Libraries, Art Institute of Chicago. On the partnership with Loebnitz

see Illinois Chapter of the American Institute of Architects, *In Memoriam Augustus Bauer*, 7.

24. Illinois Chapter of the AIA, *In Memoriam Augustus Bauer*, 6.

25. "Asher Carter" (obituary), 31.

26. Geraniotis, "German Architectural Theory and Practice," 297; and Illinois Chapter of the AIA, *In Memoriam Augustus Bauer*, 5.

27. Bryan, "The Music Hall Disaster" (letter to the editor), *Chicago Press and Tribune*, 29 November 1859. The partly constructed building had collapsed in a storm that month, prompting Bryan to defend his architects in the press.

28. John Clifford, "Asher Carter," *Chicago Tribune*, 22 January 1877, 3.

29. Illinois Chapter of the AIA, *In Memoriam Augustus Bauer*, 12.

30. Ibid., 13.

31. "The Dedication of Graceland Cemetery," *Chicago Press and Tribune*, 31 August 1860.

32. A later article, "Graceland: A Visit to Chicago's Great Burying Ground," *Chicago Morning News*, 24 August 1881, 4, notes, "The first interment was made April 13, 1860, when the body of Daniel Page Bryan, son of the owner of the cemetery, was removed from the old City Cemetery and buried in the new ground."

33. "Graceland Cemetery," *Chicago Press and Tribune*, 28 May 1860.

34. "The Dedication of Graceland Cemetery," *Chicago Press and Tribune*, 31 August 1860. See also Ware, "Memorial of William Saunders"; and the entry for Saunders in *Who Was Who in America: Historical Volume, 1607–1896*, 465.

35. "Charter of the Graceland Cemetery Company," *Chicago Tribune*, 26 February 1861; and *Charter of the Graceland Cemetery*.

36. *Charter of the Graceland Cemetery*.

37. Reynolds, *Limit of the Police Power*, 7.

38. See De Mare, *G. P. A. Healy*, 199.

39. Healy, *Reminiscences*, 62.

40. Ibid., 64.

41. Bigot, *Life of George P. A. Healy*, 31. The "real friends" whose advice resulted in profitable investments presumably included Thomas Bryan.

42. Letter from Healy to Mrs. Goddard, quoted in De Mare, *G. P. A. Healy*, 202. Madeleine Vinton Goddard (later Dahlgren), a longtime confidante of Healy's, was the daughter of Samuel F. Vinton, an Ohio congressman and leader of the Whig party.

43. Elias Olson (1815–1872) was Graceland's first superintendent; according to the cemetery's charter he was expected to keep order and "expel from the cemetery any person disturbing its sanctity by noisy, boisterous or other improper conduct." A newspaper notice by Olson, written to refute an accusation of excessive tree cutting by his workmen, offers an insight into the nature of some of his other activities: Olson wrote that he "guarded against the mutilation or cutting down of any trees except such as were in the way of monuments or other improvements," adding, "I have reason to believe that the lot owners of Graceland are satisfied with my endeavours to promote improvements and prevent the deterioration of the grounds. See "A Card from the Superintendent of Graceland Cemetery," *Chicago Tribune*, 24 January 1865, 4, responding to "Desecration of Our Rural Cemeteries" (letter, signed "A Lot Owner"), *Chicago Tribune*, 23 January 1865, 4. Olson (sometimes spelled Olsen) is buried at Graceland, in a family lot he purchased in Section A.

44. "Our Burial Places," *Chicago Tribune*, 11 December 1862, 4. Saunders is misidentified in the article as Edgar Saunders, who was perhaps better

known to the reporter as he was a local nurseryman and horticulturalist.

45. Reynolds, *Limit of the Police Power*, 7.

46. "Cemeteries. Graceland," in Spencer and Griswold, *W. S. Spencer's Chicago Business Directory*, xxiv–xxvvi; quotation on xxv.

47. "Burying the Dead: Graceland Cemetery," *Chicago Tribune*, 17 July 1864, 4.

4. A Decade of Expansion

1. Johnson and Malone, *Dictionary of American Biography*, 2:190–91.

2. *National Cyclopaedia of American Biography*, 3:170.

3. Andreas, *History of Chicago*, 2:310. The structure served "to provide for sick, wounded and destitute soldiers, and to furnish all with refreshments and temporary lodging gratuitously."

4. Johnson and Malone, *Dictionary of American Biography*, 2:190–91. The original of the document was later lost in the Great Fire.

5. *Memorial Book of the Old Tippecanoe Club*, 36–37. The Military Order of the Loyal Legion of the United States was founded in 1865 as a patriotic order of presidents and military personnel who distinguished themselves serving the Union. Bryan's election to its ranks was a signal honor, as comparatively few civilians were elected to membership. According to Keith G. Harrison, now Commander-in-Chief of the organization, Bryan was made a 3rd Class Companion sometime between 1886 and 1888. At that time this rank was reserved for "gentlemen who had remained in civil occupation during the war, but who had made measurable contributions towards the Union war effort," and it included several wartime Cabinet members and state governors. Keith G. Harrison, e-mail to author, 3 December 2008.

6. Karamanski, "Civil War"; see also his *Rally 'Round the Flag*.

7. Karamanski, "Civil War."

8. "At Graceland," *Chicago Tribune*, 31 May 1869, 1. According to this report, about a hundred Civil War soldiers were buried in the cemetery. See also "Burying the Dead," *Chicago Tribune*, 17 July 1864, 4.

9. "Our Two Cemeteries," *Chicago Tribune*, 5 December 1862, 4.

10. Bryan, "Graceland Cemetery" (letter to the editor), *Chicago Tribune*, 9 December 1862, 4. See also Bryan, "Our Two Cemeteries: 'Now, What Says Graceland?'" (letter to the editor), *Chicago Tribune*, 8 December 1862, 4.

11. "An Act to amend an Act entitled an 'Act to incorporate the Graceland Cemetery Company' approved February 22d 1861, and to incorporate the Trustees of the Graceland Cemetery Improvement Fund," approved 16 February 1865, Corporation Records, Illinois State Archives (hereafter cited as "Act to Amend"). It was also printed in *Private Laws of the State of Illinois . . . 1865*, 1:222–25. I thank John Reinhardt, supervisor of the Accessions and Control/Reference sections at the Illinois State Archives, for his invaluable assistance in locating the corporation records of the Graceland Cemetery Company. See also "Illinois Legislature," *Chicago Tribune*, 2 February 1865, 1; and "Our Cemeteries," *Chicago Tribune*, 25 June 1865, 4. See also *Journal of the Senate of the Twenty-Fourth General Assembly of the State of Illinois . . . January 2, 1865*, 324, 356, 404, 405, 497, 498, 840, 841, 944, 947; and *Journal of the House of Representatives of the Twenty-Fourth General Assembly of the State of Illinois . . . January 2, 1865*, 712, 728, 790, 819, 820, 1070, 1071. I thank Jane Ehrenhart, supervisor of Reference & Technical Services at the Abraham Lincoln Presidential Library and Museum in Springfield, Ill., for her expert assistance with this documentation.

12. "Act to Amend," 3. On the election of the trustees by lot owners, see *Catalogue of the Graceland Cemetery Lot Owners*, 7.

13. Andreas, *History of Chicago*, 2:448.

14. Sniderman, Ryckbosch, and Taylor, National Register of Historic Places Nomination Form: Lincoln Park, sec. 8, 50.

15. "J. F. G.," "Cemetery Park," *Chicago Tribune*, 29 June 1864, 4.

16. Sniderman, Ryckbosch, and Taylor, Lincoln Park Nomination Form, sec. 8, 50.

17. "The Cemetery Park," *Chicago Tribune*, 27 March 1864, 4.

18. "J. F. G.," "Cemetery Park," *Chicago Tribune*, 29 June 1864, 4.

19. See Sniderman, Ryckbosch, and Taylor, Lincoln Park Nomination Form, sec. 8, 50. See also Currey, *Chicago: Its History and Its Builders*, 167.

20. Whether Nelson drew the plan himself is unclear. The original drawing, which is held by the Chicago History Museum, is titled "Plan of Lincoln Park Chicago Adopted by the Board of Public Works" at the upper left corner, and in the lower right corner are the words, "Designed by Swain Nelson Landscape Gardener."

21. For the council's description of the site and suggested park uses, see Sniderman, Ryckbosch, and Taylor, Lincoln Park Nomination Form, sec. 8, 49, quoting Chicago Board of Public Works, *Annual Reports of the Board of Public Works to the Common Council of the City of Chicago, 1862–1869*.

22. Sniderman, Ryckbosch, and Taylor, Lincoln Park Nomination Form, sec. 7, 4–5.

23. See Maloney, *Chicago Gardens*, 374.

24. "Lincoln Park," *Chicago Tribune*, 10 June 1867, 4.

25. Reynolds, *Limit of the Police Power*, 7. See also "From Springfield: Graceland Cemetery Bill," *Chicago Tribune*, 27 January 1867, 1; and "The State Legislatures . . . Graceland Cemetery Extension," *Chicago Tribune*, 28 January 1867, 2.

26. On the town's actions see [Graceland Cemetery Company], *Proposition*, 2.

27. Reynolds, *Limit of the Police Power*, 17.

28. "The Town of Lake View v. Frederick Letz et al.," in Freeman, *Reports of Cases at Law*, 81–85; quotation on 83. See also Reynolds, *Limit of the Police Power*, 17.

29. Samuel S. Greeley, *Plat of Subdivision of 1 to 22 inclusive of Graceland Cemetery* (27 August 1868), Graceland Cemetery, Deed Box, no 1.

30. "At Graceland," *Chicago Tribune*, 31 May 1869, 1.

31. "The Cemeteries of Chicago," *Chicago Tribune*, 27 June 1869, 3.

32. Reynolds, *Limit of the Police Power*, 7.

33. [Graceland Cemetery Company], *Proposition*, 2. See also Waller, *Right of Eminent Domain;* Waller dates the ordinance to 5 March 1867.

34. Saunders's correspondence register records a letter to Bryan on 21 February 1870, and one received from him on 26 February. William Saunders Papers, National Grange of the Patrons of Husbandry Records, Cornell University Library, box 1, [Correspondence] register 2.

35. Important studies of Cleveland and his work include Hubbard, "H. W. S. Cleveland"; Nadenicek, "Nature in the City"; Tishler, "Horace Cleveland: The Chicago Years"; and Daniel J. Nadenicek and Lance M. Neckar's introduction to the Library of American Landscape History reprint of *Landscape Architecture as Applied to the Wants of the West*.

36. See Nadenicek, "Sleepy Hollow Cemetery: Philosophy Made Substance" and "Sleepy Hollow Cemetery: Transcendental Garden and Community Park."

37. Nadenicek and Neckar, "Introduction," xx.

38. Ibid., xxi.

39. See Gerdts, "Chicago Is Rushing Past Everything." Lakey was apparently wrong about Chicago, however; around 1873, *American Builder* began to be published in New York.

40. "Brooklyn Park Improvements," 71.

41. Rauch, "Public Parks."

42. "The Architect and the Landscape Gardener," *American Builder and Journal of Art* 2, no. 5 (May 1869): 100.

43. "Personal," *American Builder and Journal of Art* 2, no. 5 (May 1869): 116.

44. H. W. S. Cleveland, "A Few Hints on the Arrangement of Cemeteries," *Chicago Tribune*, 18 October 1869, 1. All quotations are taken from this article.

45. *American Builder and Journal of Art* 3, no. 7 (July 1870): 167. An article by French, "Country Roads," appeared in the same issue.

46. Writing for *American Builder* also linked the three men and suggests that at the very least they thought highly of and were well known to one another. Greeley, for instance, in an article titled "The Laying Out of Western Suburban Villages," lauded Cleveland's roads as exemplars of the "curved" or "landscape method," and a few months later French praised Greeley's piece in an article on the same topic, "Methods of Surveying Irregular Sub-divisions."

47. All quotations in this paragraph are from "The Improvement of Lots" in [Graceland Cemetery Company], *Catalogue of the Graceland Cemetery Lot Owners*, 10.

48. French, "Trees in Composition," 59–60.

49. Bryan Lathrop to J. S. Birkeland, 25 March 1878, Graceland Cemetery File Book, p. 18, Cemetery Office, Graceland. On Rascher see Gemperle, "The Raschers of Andersonville"; and Ristow, *American Maps and Mapmakers*. For an overview of the mapmaking field in Chicago at this time, see Conzen, "Evolution of the Chicago Map Trade."

50. Three copies of Rascher's map are known to exist; all are in Chicago, two at the Chicago History Museum and the third at the Newberry Library.

51. By the time Rascher mapped the cemetery, he had become known for the production of documentary plans such as his *Fire Insurance Map of Chicago* (1877), the first atlas of its type to appear in the city. This has led some to presume that he similarly recorded Graceland as it actually then existed. If he had made the map for purely documentary purposes, however, producing it in color and including elaborate elevations of individual trees and shrubs would hardly have been necessary. These attributes suggest that the image had dual purpose.

52. This tract, as we will see, was formerly the site of Lake View pioneer Conrad Sulzer's homestead, and the comparatively luxuriant, picturesque parkland pictured on Rascher's map was Sulzer's own handiwork. This landscape vignette also enables us to more precisely fix the date of the map itself. After Graceland purchased the property in April 1879, the house would be relocated across Green Bay Road to the opposite corner; thus its absence from Rascher's view suggests his map dates to that year. We know from Bryan Lathrop's letter to Birkeland (instructing him to assist Rascher) that the cartographer began his project around March 1878. After the cartography had been accomplished, maps such as this one could be produced in a matter of months. Unless his survey was delayed, the interim period between March 1878 and April 1879 was more than adequate to produce the map. This circumstance, in turn, raises the possibility that his image anticipated or projected the moving of the house. See Reps, *Views and Viewmakers*, 45–52.

53. On the "miniature lakes" see Reynolds, *Limit of the Police Power*, 41.

54. Ibid., 6. Spring Grove was chartered in 1845 by the Cincinnati Horticultural Society, which aimed to follow Mount Auburn's lead. John Notman (1810–1865), the designer of Laurel Hill, was commissioned to lay out the 166-acre cemetery that same year, but his plan was set aside, as was that of the New York landscape gardener Howard Daniels (1815–1863) the next year, and development apparently languished until Strauch arrived in the winter of 1854–55 "to implement his plan of improvement and assume the position of superintendent and landscape gardener." See Blanche M. G. Linden's entry on Notman and Christine B. Lozner's on Daniel in Birnbaum and Karson, *Pioneers of American Landscape Design*, 269–72 and 73–76.

55. See, for example, N. D. Vernon, "Adolph Strauch: Cincinnati and the Legacy of Spring Grove Cemetery." Strauch, a Prussian émigré who had apprenticed in the imperial gardens in Vienna and studied in Paris, came to the United States in 1851 after a stint in London at the Royal Botanic Society gardens in Regent's Park. He settled in Cincinnati, then a horticultural center, the next year and established a private landscape gardening practice. Strauch was strongly influenced by Hermann Ludwig Heinrich, Prince von Pückler-Muskau (1785–1871), an exponent of picturesque landscape design and the author of *Hints on Landscape Gardening* (1834). Working primarily on large private estates, Strauch took the English picturesque, as filtered through Pückler-Muskau's writings, as his primary aesthetic source.

56. Reynolds, *Limit of the Police Power*, 6.

57. Creese, "Graceland Cemetery, 209.

58. N. D. Vernon, "Strauch, Adolph," in Birnbaum and Karson, *Pioneers of American Landscape Design*, 385.

59. N. D. Vernon, "Adolph Strauch: Cincinnati and the Legacy of Spring Grove Cemetery," 14.

60. Kirkland, *History of Chicago*, 2:604.

61. Andreas, *History of Chicago*, 2:449.

62. For about a year into Cleveland's Chicago practice, *American Builder and Journal of Art* continued to publish his essays and to promote his work with lavish praise; in May 1870 it wrote of the "great success" that had come to him in the city and called his talent "of the silver-plated order" (120). In June 1871 the magazine's editor cautioned those in need of a landscape architect against hiring a "quack"; rather, he would "very cordially commend Mr. H. W. S. Cleveland" (433). Other evidence, however, raises the possibility that Cleveland's work at Graceland might not have gone well. The 1870 *Catalogue of the Graceland Cemetery Lot Owners* is the only known surviving source to definitively document Cleveland as one of the cemetery's authors. Cleveland himself apparently did not cite his Graceland work in any of his own writings. Why would he omit reference to this substantial project, when in his own advertisements in *American Builder*, such as the one that appeared in the March–April 1871 issue (iii), he named John J. Smith of Philadelphia's Laurel Hill Cemetery as a reference? He also included local Chicagoans among those endorsing his talents, yet the names of anyone connected to Graceland are absent from his list of references. Surely he would not omit the name of such a well-known local figure as Thomas Bryan if he thought Bryan willing to vouch for him. And, as we will see, when next again in need of a landscape gardener, the cemetery would not return to Cleveland.

63. Creese, "Graceland Cemetery," 209.

64. Waller, *Right of Eminent Domain*, 16.

65. Creese, "Graceland Cemetery," 209.

66. Waller, *Right of Eminent Domain*, 12.

67. Reynolds, *Limit of the Police Power*, 9.

68. All quotations in this paragraph ibid., 40–41.

69. Reynolds noted, "This pamphlet was first written eight months ago; and the manuscript was ready for the printer, but it was destroyed in the fire of October 9th." Ibid., 3.

70. Cook, *Bygone Days in Chicago*, 178.

71. Berens, *Elmhurst*, 89. Berens notes: "The panic of 1873 brought further losses; it is said that fire and panic cost Bryan the major part of a $2,000,000 fortune. But, bravely he joined in rebuilding the city and courageously he set to work to recoup his financial losses. Of the great fire refugees given shelter by the Bryans and their village neighbors, a considerable number bought Elmhurst property and built homes, thus initiating a second period of growth for the suburb."

72. "Gallery of Local Celebrities. No. XVI. Thomas B. Bryan," *Chicago Tribune*, 13 May 1900, 39. The *Tribune* estimated Bryan's total losses from the fire at more than $500,000.

73. Geo. H. Frost, Surveyor, *Map of Section M. Graceland Cemetery* (4 December 1872), Graceland Cemetery Collection (1994.146), Series I: Maps and Plats, map folder 3, Chicago History Museum.

74. "The Town of Lake View v. The Rose Hill Cemetery Company [September Term, 1873]," in Freeman, *Reports of Cases at Law*, 191–204; quotation on 195–96.

75. Ibid., 201.

76. Record Book, Graceland Cemetery Co. (1873–1937), 17 and 19 (15 September 1874), Cemetery Office, Graceland.

5. Bryan Lathrop and William Le Baron Jenney

1. Thomas Patterson replaced Bryan as president of the Graceland Cemetery Company, serving until Bryan Lathrop became president in 1881. Although the precise date Bryan left Chicago for Washington is unclear, President Hayes sent his nomination to the Senate on 17 October 1877. See "Nominations by the President," *New York Times*, 18 October 1877. On Bryan's time in Washington, see Green, *Washington*, 383–87.

2. Later that summer, however, the depression would be prolonged by the Great Railroad Strike of 1877.

3. On the ruling see "The Law Courts," *Chicago Tribune*, 13 February 1878, 3, and "The Law: A Little Tax Law," *Chicago Tribune*, 23 November 1879, 13.

4. Classified advertisement, *Chicago Tribune*, 10 April 1877, 1.

5. The advertisement appeared in the *Chicago Tribune* on the following dates: 15 and 17 April 1877, 6 and 27 May 1877, 17 and 24 June 1877, and 1 July 1877. These apparently were the only display advertisements Graceland ran that year, and none appeared in 1878.

6. Record Book, Graceland Cemetery Co. (1873–1937), 27 (23 June 1877), Cemetery Office, Graceland (hereafter cited as Record Book, Graceland).

7. Reynolds, *In the Supreme Court of Illinois*, 3.

8. Creese, *Crowning of the American Landscape*, 233.

9. Untitled, *American Builder and Journal of Art* 1, no. 1 (15 October 1868): 7–8; see also "Description of the Plates," 9. The houses illustrated are "Residence of John Forsythe Esq. Douglas Grove, Chicago" and "Cheap Suburban Residence, Chicago."

10. Creese, *Crowning of the American Landscape*, 232.

11. Loring and Jenney, *Principles and Practice of Architecture*, 32.

12. *American Builder and Journal of Art* 3 (February 1870): 44.

13. "Residence of Watts DeGolyer," *American Builder and Journal of Art* 3 (November 1870): 253.

14. "Residence of Charles Gladding, Esq., Riverside," *American Builder and Journal of Art* 3 (December 1870): 274.

15. Greeley, "Laying Out of Western Suburban Villages," 371.

16. Rainey, "William Le Baron Jenney and Chicago's West Parks," 58.

17. For a brief overview of Haussmann's role in implementing Napoleon III's plans for the "modernization and security" of Paris, see Denis Lambin's entry on Haussmann in Goode and Lancaster, *Oxford Companion to Gardens*, 246.

18. *Biographical Dictionary and Portrait Gallery*, 39.

19. Marietta and Cincinnati Railroad Company, *Map of rail road line between Loveland and Cincinnati; Marietta and Cincinnati Rail Road, . . . Reorganized August 1, 1860.* A copy is held at the Library of Congress; see http://hdl.loc.gov/loc.gmd/g4083h.rr004550.

20. *Biographical Dictionary and Portrait Gallery*, 39.

21. Rainey, "William Le Baron Jenney and Chicago's West Parks," 58.

22. Jenney to Olmsted, 16 December 1865, Papers of Frederick Law Olmsted, Manuscript Division, Library of Congress, microfilm reel 9, frames 561–63.

23. Jenney to Olmsted, 2 December 1865, ibid., frames 546–48.

24. All quotations in this paragraph are from *Biographical Dictionary and Portrait Gallery*, 39.

25. The partners are listed as "L. Y. Schermerhorn, John Bogart and L. Y. Colyer"; the third is perhaps John Y. Culyer, who had been the chief engineer for Prospect Park in Brooklyn. See the entries on Bogart and Culyer by Joy Kestenbaum in Birnbaum and Foell, *Shaping the American Landscape*.

26. *Book of Chicagoans*, 348. O. C. Simonds later wrote: "When [Bryan Lathrop] had prepared for college at his home in Virginia, his father gave him the choice of four years in the University of Virginia or four years in Europe. He chose the latter and spent two years living in the house of a German professor, and most of the remaining time in France." Simonds, "Notes on Graceland," 8.

27. Simonds noted, "He visited and studied the art galleries of Europe, comparing his own judgement of the various paintings with that of experts," adding that Lathrop "might, indeed, have been called a critic in all of the fine arts, but probably landscaping was his favorite among them all." "Notes on Graceland," 8.

28. Lathrop, "Parks and Landscape-Gardening," 8.

29. Simonds, "Notes on Graceland," 8–9.

30. All quotations in this paragraph describing the details of the work at Huntington are from Bryan Lathrop to John F. Cremin, 22 September 1897, Elmhurst Historical Museum.

31. Simonds, "Notes on Graceland," 8. Huntington was not the only Cottage Hill country place under development in the 1860s. Throughout this period, for instance, Thomas Bryan continued to subdivide his land holdings there, and in 1865 he sold a portion of his property to Seth Wadhams (1812–1888), a Chicago ice magnate. Wadhams began cultivating his new estate, called White Birch, and employed Swain Nelson to assist him with the task. Bryan also sold parcels that became St. Mary's Cemetery and a German Evangelical seminary (today Elmhurst College), and it is possible that he recommended Nelson, already working at Cottage Hill, as landscape gardener for these sites. By 1869, after a decade's residence at Bird's Nest, Bryan's own landscape gardening activity was undiminished. An

Elmhurst pioneer recollected Bryan and his neighbors Lathrop and Wadhams as having "spared no expense, or effort, to obtain the finest specimens of all the trees and shrubs that would grow in this latitude, developing them to the greatest degree of beauty with the aid of skilled gardeners." Following Bryan's lead, the village had by now evolved into a network of at least ten palatial estates, winning renown as a garden spot. See, for instance, "How Chicago Suburbs Were Planted and Named," *Chicago Tribune*, 25 February 1900, 33; and Berens, *Elmhurst*, 103–18.

32. Record Book, Graceland, 28 (23 October 1878).

33. Simonds, "Notes on Graceland," 8.

34. Ibid.

35. Ibid., 9. On Downing's writings, see Major, *To Live in the New World*. Also invaluable are the essays in Tatum and MacDougall, *Prophet with Honor*.

36. Simonds, "Notes on Graceland," 9.

37. Ibid. Simonds did not supply Gilpin's first name, and it is unclear whether he was referring to the painter and aesthetic theorist William Gilpin (1724–1804) or to William Sawrey Gilpin (1762–1843). Both men authored works germane to landscape gardening: William Gilpin's *Three Essays: On Picturesque Beauty; On Picturesque Travel; and On Sketching Landscape* (1792) and William Sawrey Gilpin's well-known *Practical Hints upon Landscape Gardening* (1832).

38. Robinson, "Garden Cemeteries." Robinson actually wrote only the article's introductory paragraph, extracting the remaining text from one of the cemetery's reports, identified only as "the published report of Spring Grove Cemetery" (186). Among Robinson's other writings on cemeteries are *God's Acre Beautiful; or, The Cemeteries of the Future* (1883). General works on Robinson include Allan, *William Robinson, 1838–1935*, and Bisgrove, *William Robinson*.

39. Robinson, "Garden Cemeteries," 186.

40. Simonds, "Notes on Graceland," 9. Simonds wrote "Cleveland French," apparently mistaking the two for a single person. But he characterized "Cleveland French" as the one "who as late as 1880 had the only landscape gardening office in Chicago." That Simonds remembered both Cleveland and French in association with Bryan Lathrop raises the possibility that Cleveland's partner may also have had a hand in Graceland's layout.

41. All quotations in this paragraph are from Lathrop, "Parks and Landscape Gardening," 8–10.

42. Lathrop to J. S. Birkeland, 22 October 1877, Graceland Cemetery File Book, p. 159, Cemetery Office, Graceland.

43. Remaining quotations in the paragraph are from Simonds, "Notes on Graceland," 9.

44. Writing around 1928, H. J. Reich, then Graceland's superintendent, noted that the "plan of Section O with larger lots was due to Mr. Lathrop." [H. J. Reich], "Graceland Cemetery" (four-page unpublished typescript, ca. 1928), Cemetery Office, Graceland, 2. Charles Rascher's map of around 1878 confirms that the burial plots in Section O, in comparison to those in nearby sections, are indeed large. The 1884 *Map of Graceland Cemetery, Cook Co., Ill.* shows the identical subdivision of that section, so the project can be confidently dated to around 1878. Rascher's image also represents a pair of monuments adorning Section O, indicating that burials had already been made there, presumably after the section had been subdivided. This too confirms Lathrop's sole authorship.

45. Ranney, "Frederick Law Olmsted," 45. Although initially envisaged as a self-sufficient town, Riverside would never realize this aim.

46. Riverside Improvement Company, *Riverside in 1871*, 5.

47. Creese, *Crowning of the American Landscape*, 227.

48. Turak, *William Le Baron Jenney*, 95.

49. Ibid.

50. Creese, *Crowning of the American Landscape*, 223–24.

51. Ibid., 223.

52. Rainey, "William Le Baron Jenney and Chicago's West Parks," 62.

53. Turak, *William Le Baron Jenney*, 78.

54. Rainey, "William Le Baron Jenney and Chicago's West Parks," 62.

55. Ibid.

56. Turak, *William Le Baron Jenney*, 84.

57. "Moline," *Rock Island Union*, 20 March 1872, n.p. In 1837 John Deere became one of the first to build a commercially successful self-scouring, cast-steel plow, hastening the agricultural transformation of the prairies and ultimately fuelling Chicago's, and the nation's, growth. Indeed, by the 1860s and 1870s Moline had boomed, supplanting Pittsburgh as America's plow manufacturing capital. On Deere see Dahlstrom and Dahlstrom, *The John Deere Story*.

58. "Moline," *Rock Island Union*, 20 March 1872, n.p.

59. Photograph, "Taken from northeast showing *Overlook*, gazebo, greenhouse and carriage house" (1987.128.1), and stereograph, "Exterior view of *Overlook*—west side" (1987.125.1), William Butterworth Center & Deere-Wiman House, Moline, Ill.

60. Dahlstrom and Dahlstrom, *John Deere Story*, 101.

61. *Holland's Moline Directory*, 329. As the switch to a name more evocative of the cemetery's setting suggests, the rural cemetery movement had found its way further west. By 1855, for instance, the nearby city of Rock Island produced its own example of the type, Chippiannock Cemetery—one of the first rural cemeteries in the region, founded even in advance of Chicago's Rosehill. See *Historic Rock Island County*, 136. Chippiannock was designed by Almerin Hotchkiss (1816–1903), who also also laid out the Chicago suburb of Lake Forest, including locating its cemetery (see chapter 6).

62. "Special Meeting of the City Council," *Moline Weekly Review*, 20 February 1874, n.p.; and *Holland's Moline Directory*, 329.

63. The plan is briefly described in *Holland's Rock Island and Moline Directory*, 375; the city council meeting is reported in "City Council Proceedings," *Moline Weekly Review*, 21 August 1874, n.p.; see also "Special Meeting of the City Council August 18th 1874," in Moline City Council Minute Book, vol. 2 (1872–1883), 80 (City Hall, Moline, Ill.). The *Moline Weekly Review* described the plan as "a beautiful drawing of Riverside cemetery . . . drawn on the plan of 80 feet to the inch; show[ing] the drives, 1401 lots, to good advantage." Untitled, *Moline Weekly Review*, 21 August 1874, n.p. Two copies of Jenney's plans are known; one is held at the City of Moline Cemetery Office, Moline Memorial Park, Moline, Ill., and the other at the Rock Island County Recorder's Office, Rock Island, Ill.

64. *Holland's Moline Directory*, 329.

65. Ibid., 329.

66. Record Book, Graceland, 28 (23 October 1878). This apparently was the first meeting after the one of 23 June 1877 at which it was decided to construct the ditch and tile drain mentioned earlier. At the October meeting, the board of managers authorized Lathrop "to make a contract with J. C. Robinson for a 30-inch brick sewer to be laid on the Shippey Road from a point West of the Eastern line of the East Half of North West Quarter of

Section 17.40.14. to Lake Michigan according to the plans and specifications of W. L. B. Jenney, who is to be employed as Superintendent of the drainage and engineering and landscape gardening in Graceland under the directions of Bryan Lathrop Vice-President" (28).

67. Reynolds, *In the Supreme Court of Illinois,* 2.
68. Record Book, Graceland, 28 (23 October 1878).
69. Ibid. By now Jenney presumably had been in dialogue with the cemetery for some time, probably since as early as around April 1877, when Graceland advertised its plans to construct lakes. Indeed, although perhaps it is only a coincidence, this was the same year Jenney elected to resign his West Parks district post. As we have seen, the board's June 1877 decision to make a ditch and tile drain from the "low ground on the east side" to Lake Michigan had signaled its plans to expand east. Minutes of the board's subsequent meetings record neither the drain's designer nor whether it was actually built. Quite possibly, this "drain" and Jenney's "sewer" were one and the same project, a poor economic climate having conspired against its construction until now.
70. Writing in the third person, Simonds recollected: "In the fall of 1878 Mr. Simonds was sent from the architect, W. L. B. Jenney's office out to Graceland Cemetery. . . . He was given the job of superintending the work which was completed by the following spring [1879]." Simonds, "Dean of the Cemetery Field," 19.
71. See Turak, *William Le Baron Jenney,* 143–56, and his article "William Le Baron Jenney: Teacher."
72. Simonds, "Graceland at Chicago," 12.
73. Simonds, "Dean of the Cemetery Field," 19.
74. Ibid.

6. Final Expansion

1. [Graceland Cemetery Company], *Proposition,* 1.
2. The ordinance to limit Graceland was passed on 3 February 1879; see *Revised Ordinances of the Town of Lake View,* 133. See also "Suburban: Lake View," *Chicago Tribune,* 5 February 1879, 8.
3. [Graceland Cemetery Company], *Proposition,* 2.
4. Ibid., 1. On the meetings, see also "The City: General News," *Chicago Tribune,* 27 February 1879, 8.
5. Lake View's register of ordinances confirms that Graceland presented the town with "an application for leave to enlarge its cemetery according to the terms and conditions of the following proposition" on 10 March 1879. See *Revised Ordinances of the Town of Lake View,* 133–35. The successful conclusion of discussions between the two parties was reported in the *Tribune* the next day; see "Graceland Cemetery," *Chicago Tribune,* 11 March 1879, 8.
6. [Graceland Cemetery Company], *Proposition,* 3. The proposed new lands are described as lying east of Green Bay Road, together with eight lots in the area known as "Iglehart's Sub-division" (5–6).
7. Ibid., 6.
8. Ibid.
9. The petition is reproduced in ibid., 3–4; see also "Suburban: Lake View," *Chicago Tribune,* 5 March 1879, 8, and "Suburban: Lake View," *Chicago Tribune,* 12 March 1879, 8.
10. For the signatories see [Graceland Cemetery Company], *Proposition,* 4.
11. Ibid., 3.
12. See *Revised Ordinances of the Town of Lake View,* 135; see also "Suburban: Lake

View," *Chicago Tribune*, 9 April 1879, 8. The extension was approved by a vote of 505–81; see "Miscellaneous: Lake View," *Chicago Tribune*, 3 April 1879, 5.

13. Geiger, "Nature as the Great Teacher," 23.

14. It is not known exactly when the land was purchased from Sulzer, but it might in fact have been months earlier, in advance of the Lake View settlement. At its 27 November 1878 meeting, Graceland's board of managers had resolved "to purchase the land belonging to Mrs. Sulzer [Conrad Sulzer's widow], being five acres more or less, in the North West corner of the cemetery," and authorized the company's chief officers to arrange the terms. Record Book, Graceland Cemetery Co. (1873–1937), 31 (27 November 1878), Cemetery Office, Graceland (hereafter cited as Record Book, Graceland).

15. On Conrad Sulzer, see Bjorklund and Haglund, *Pioneer Settler*. Also useful are Drury, *Old Chicago Houses*, 233–37; and Clark, *Lake View Saga*, 16–17. See also "Lake View Hails 1st Settler in Rites Tuesday," *Chicago Tribune*, 28 May 1950, N6; and the obituary of his granddaughter Grace E. Sulzer, *Chicago Tribune*, 25 October 1957, B2. Sulzer is buried at Graceland, in Section A, and a marker in Section R records the approximate location of his homestead site.

16. Frederick Sulzer went to Rochester, N.Y., then a center of floriculture, for two years to study nursery techniques, and in 1860, a year after his return, he opened a nursery selling plants and trees and also offered his services as a landscape architect. Clark, *Lake View Saga*, 16; Bjorklund and Haglund, *Pioneer Settler*, 6. Bjorklund and Haglund write that by 1870 he "had expanded his nursery business to include a florist shop serving the expanding population of Ravenswood and Lake View as well as growing numbers of mourners at Graceland Cemetery across Green Bay Road and those bound for Calvary Cemetery in Evanston." *Pioneer Settler*, 8.

17. Biographical information and quotations in this paragraph are from "In Memoriam: John A. Cole," *United Church Chronicle*, December 1932, 5–7. A copy of this publication is available at the Sulzer Regional Library branch of the Chicago Public Library.

18. By the time of his death in 1932, Cole had served as "Elder and Elder emeritus of the Hyde Park Presbyterian Church and of the United Church of Hyde Park for over fifty years." On the U.S. Christian Commission and Cole's role in it, see Cannon, "United States Christian Commission." Cannon characterizes the organization as "sister" to the United States Sanitary Commission.

19. While living in Washington, Cole also served on the board of trustees of Howard University. See Logan, *Howard University*, 52, and Dyson, *Howard University*, 19–20, 50, 408, 414, 419. I thank Dr. Clifford L. Muse Jr., Howard University archivist, for supplying me with copies of this material.

20. "In Memoriam: John A. Cole," 6. On Cole and waterworks see Keating, *Building Chicago*, 92, 93, 146n27.

21. "In Memoriam: John A. Cole," 6.

22. Keating, *Building Chicago*, 146n27.

23. Cole raised the waterworks' crib and excavated around the inlet pipe to stop sand from flowing in. See "Suburban: Lake View," *Chicago Tribune*, 1 December 1878, 7.

24. "Ravenswood: Street Improvements," *Chicago Tribune*, 5 September 1875, 7. Also that year, Cole erected a building in Ravenswood, presumably his office, "on the corner of Commercial Street and Cosgrove Avenue." See "Ravenswood," *Chicago Tribune*, 10 January 1875, 2.

25. John Drury, writing in 1941, noted that "ornamental shrubs and trees of the

original Sulzer place" could still be found growing in Section R. Drury, *Old Chicago Houses*, 234.

26. John A. Cole, "Cross-Sections in Graceland Cemetery" (March 1879), Graceland Cemetery Collection (1994.146). Series I: Maps and Plats, map folder no. 5.

27. N. P. Iglehart was a developer and real estate agent, well known to Thomas Bryan. In the 1860s, for instance, they served as officers for Chicago's Board of Real Estate Brokers; see "60 Years Ago Today," *Chicago Tribune*, 2 August 1925.

28. *A Plat Showing the Lands of the Graceland Cemetery Company and the proposed final boundaries of Graceland Cemetery* was a foldout map accompanying *Proposition of the Graceland Cemetery Company*. According to this map, the cemetery also owned Iglehart lots immediately to the east of its proposed eastern boundary: lots 2, 5, 6, 9, and 18. The purchase dates of these earlier acquisitions are unclear. See also Record Book, Graceland, 37 (30 April 1879).

29. At its 9 May 1879 meeting, the board of managers accepted the plans Jenney had prepared, ordering the roads he laid out to be built and voting to "make contracts for the excavation and removal of soil according to the plans and specifications of Mr. Jenney." Record Book, Graceland, 38.

30. A copy of this map is held by the Chicago History Museum (ICHi 31174).

31. On the lotus and its symbolic resonances with the afterlife, see Griffiths, *Lotus Quest*.

32. Simonds, "Dean of the Cemetery Field," 19.

33. Ibid.

34. See Turak, *William Le Baron Jenney*, 101–3 (illustration on 102).

35. Simonds, "Dean of the Cemetery Field," 19.

36. Simonds's successor as superintendent, H. J. Reich, recorded that Graceland's earthworks were made from, among other sources, "material from the excavation for lakes." [H. J. Reich], "Graceland Cemetery" (four-page unpublished typescript, ca. 1928), Cemetery Office, Graceland, 2. Simonds himself recollected that "during 1879 this and the ground below it, sand and gravel, were excavated to form a lake [Willowmere]." Simonds, "Dean of the Cemetery Field," 19; see also Simonds, "Graceland at Chicago," 12.

37. On the French picturesque see, for instance, Hunt, *Picturesque Garden in Europe*, 90–103 and 104–39.

38. All quotations in this paragraph are from Turak, *William Le Baron Jenney*, 63.

39. Ibid., 62.

40. Robinson, *Parks, Promenades, & Gardens of Paris*, 66–78; quotation on 66.

41. On Ermenonville see Hunt, *Picturesque Garden in Europe*, 117–19.

42. Creese, *Crowning of the American Landscape*, 220.

43. Ibid., 227.

44. Turak, *William Le Baron Jenney*, 69.

45. Remaining quotations in this paragraph are from Turak, *William Le Baron Jenney*, 107–8.

46. André, *L'art des jardins*, 2:873.

47. Ibid., 2:868.

48. See Alaimo, National Register of Historic Places Nomination Form: Lake Forest Cemetery.

49. Coventry, Meyer, and Miller, *Classic Country Estates of Lake Forest*, 38. On Hotchkiss see the entry by Arthur H. Miller in Birnbaum and Foell, *Shaping the American Landscape*.

50. Alaimo, Lake Forest Cemetery Nomination Form, section 8, 16. Alaimo

identifies the local engineer as Samuel F. Miller, who was a civil engineer with the Chicago and Milwaukee Railroad and also the first principal of Lake Forest Academy.

51. Ibid., section 8, 18.
52. Rainey, "William Le Baron Jenney and Chicago's West Parks," 67.
53. On French influence in the West Parks, see Rainey, "William Le Baron Jenney and Chicago's West Parks."
54. Jenney's "Plan and Setting, Winter Garden, Douglas Park, Chicago, 1888" is reproduced in Turak, *William Le Baron Jenney*, 234.

7. The Era of Bryan Lathrop and O. C. Simonds

1. Record Book, Graceland Cemetery Co. (1873–1937), 49 (9 January 1880), Cemetery Office, Graceland (hereafter cited as Record Book, Graceland).
2. Ibid.
3. Edward Renwick manuscript (1932), unpublished memoir, Northwest Architectural Archives, Manuscripts Division, University of Minnesota, Minneapolis, 28–29 (hereafter cited as Renwick Manuscript).
4. See Bruegmann, *Architects and the City*, 3–15. According to Bruegmann, Jenney "became a kind of father figure to a whole generation of prominent architects that flourished in the city in the last two decades of the nineteenth century" (9).
5. Jenkins, "Review of the Work of Holabird & Roche," 1. This source unfortunately does not identify the specific date of Holabird's departure. It is possible, but not probable, that he left Jenney's employ earlier than did Simonds.
6. See Bruegmann, *Architects and the City*, 3–15. Bruegmann writes: "It seems likely, then, that Simonds was somewhat more important than Holabird in creating the firm and securing work that attracted attention. Twenty-five years old at the time the firm was founded, he was a year younger than Holabird, but he had completed his university education, whereas Holabird had not, and his Graceland Cemetery connection provided the partners with important potential commissions and valuable contacts" (10).
7. Edward Renwick recollected that Jenney "turned this work [Graceland] over to Mr. Simonds." Renwick Manuscript, 28.
8. See Bruegmann, *Architects and the City*, 7–10.
9. For Lathrop's election see Record Book, Graceland, 51 (1 October 1881); for Simonds's official title see 57 (16 January 1885).
10. Simonds, "Notes on Graceland," 9.
11. Simonds, "Graceland at Chicago," 16.
12. Ibid., 12. Similarly, one of Simonds's successors believed that Graceland's beauty was "due, in very large measure, to the exquisite taste of Bryan Lathrop." See [H. J. Reich], "Graceland Cemetery" (four-page unpublished typescript, ca. 1928), Cemetery Office, Graceland, 2 (herafter cited as [Reich], "Graceland Cemetery").
13. Lathrop, "A Plea for Landscape Gardening," 324.
14. Significant studies of Simonds include Grese, "Prairie Gardens of O. C. Simonds and Jens Jensen," and his introduction to the Library of American Landscape History reprint of Simonds's *Landscape-Gardening;* and Bachrach, "Ossian Cole Simonds: Conservation Ethic in the Prairie Style."
15. University of Michigan, *Calendar*, 60.
16. On Jenney and the University of Michigan architecture course see Turak, *William Le Baron Jenney*, 143–49, and his article "William Le Baron Jenney: Teacher."

17. "Ossian Cole Simonds" (*American Landscape Architect* obituary), 17.

18. Registered copy of "Simonds, Ossian Cole" transcript, provided by the Office of the Registrar, University of Michigan, Ann Arbor.

19. University of Michigan, *General Register*, 70–71.

20. Ibid., 71. Also included in the description are "Use of instruments. Topography. One Plane Descriptive. Setting out work. Plane Table."

21. Ibid.

22. "From lack of appropriations, the department of architecture was abandoned and Professor Jenney returned to Chicago." "Ossian Cole Simonds" (*American Landscape Architect* obituary), 17.

23. Simonds was joined on the survey by John B. Johnson, a classmate and friend. See Leland and Smith, "Ossian Cole Simonds: Master of Landscape Architecture." Barbara Geiger reports that Simonds had also worked for the U.S. Geodetic Survey during the summers before his graduation; see Geiger, "Nature as the Great Teacher," 16.

24. "Ossian Cole Simonds" (*American Landscape Architect* obituary), 17.

25. Pond, *Autobiography*, 157.

26. The precise date that Simonds and Arey began lodging together is unclear. Arey, after graduating in 1878, started his career in Buffalo, N.Y., "but within a year entered Professor Jenney's office in Chicago." T. R. Chase, "Clarence Oliver Arey, Architect and Physician," included in Arey's necrology file, Bentley Historical Library, University of Michigan. According to other documents in the file, Chase obtained Arey's biographical details from the architect's widow. I thank Marilyn McNitt of the Bentley Historical Library for supplying me with photocopies of this document.

27. Pond, *Autobiography*, 157.

28. In one of his historical accounts on Graceland, Simonds, for instance, remembered Jenney only as "the architect who had drawn an outline of the lake and planned its outlet." See Simonds, "Graceland at Chicago," 12.

29. "Through Mr. Lathrop's influence he became interested in landscaping." See "Ossian Cole Simonds" (*American Landscape Architect* obituary), 17 .

30. Pond, *Autobiography*, 158.

31. Simonds, "Graceland at Chicago," 12.

32. Ibid.

33. Pond, *Autobiography*, 158.

34. Writing in the third person, Simonds recollected that it "did not please him to be working in Graceland when he had planned and wished to be working at architecture. Simonds, "Dean of the Cemetery Field," 21.

35. Pond wrote, "I may say now that Simonds eventually supplanted Major Jenney and became not only Landscape builder for Graceland Cemetery but became also a national figure in the profession into which he had as it were been shunted." *Autobiography*, 158.

36. Simonds, "Dean of the Cemetery Field," 21. Another source enlarges: "With his brother he soon began to identify and catalogue the plants and flowers of the countryside. This childhood love of Nature, and this interest in botany, matured as the boy grew into manhood." Leland and Smith, "Ossian Cole Simonds: Master of Landscape Architecture." His obituary in *American Landscape Architect*, too, notes that he "often spoke of the youth spent with his brother in collecting and identifying every flower, shrub and tree on the farm and in the neighborhood" (17).

37. On Jenney's prominence within Chicago's professional scene, see, for example, David Van Zanten's insightful account in "Sullivan to 1890."

38. Turak cites Victor Ruprich-Robert's *Flore ornementale* (Paris, 1866) and

James Colling's *Art Foliage* (London, 1865) as among Jenney's "sources for his ornamental art"; see Turak's *William Le Baron Jenney*, 122, and his article "The École Centrale and Modern Architecture." Turak discusses the theoretical underpinnings of Jenney's work in *William Le Baron Jenney*, 113–56.

39. See Egbert and Sprague, "In Search of John Edelmann"; and Mallgrave, *Modern Architectural Theory*, 163–64.

40. Menocal, "Iconography of Architecture," 83. Menocal's study is an outstanding source on Sullivan's time with Jenney. Sullivan is thought to have joined Dankmar Adler's firm by 1880.

41. Pond, *Autobiography*, 158.

42. See, for instance, Narciso G. Menocal's brilliant monograph, *Architecture as Nature*.

43. Pond, *Autobiography*, 159.

44. On Sullivan and Edelmann see Van Zanten, "Sullivan to 1890"; see also Twombly, *Louis Sullivan*, 50–52, 70–71, 82–87.

45. Arey apparently remained in Jenney's office until 1883, when he moved and established his own architectural practice in Cleveland, Ohio. Although his works were primarily residential, in 1888 he rebuilt the Main Building for the Case School of Applied Science (now Case Western Reserve University), which had burned two years earlier. He also designed Adelbert Gymnasium for Western Reserve (Adelbert College). After losing his only son to diphtheria in 1891, Arey, poignantly and quite remarkably, left architecture for medicine and graduated with an M.D. from the University of Pennsylvania in 1894. He died two years later from "overwork." T. R. Chase, "Clarence Oliver Arey, Architect and Physician," included in Arey's necrology file, Bentley Historical Library, University of Michigan. I thank Helen Conger, Case Western Reserve University Archives, Cleveland, Ohio, for providing information on Arey's work there.

46. On Pond and the Prairie School, see H. Allen Brooks's classic study, *The Prairie School*.

47. All quotations in this paragraph are from Lathrop, "Parks and Landscape-Gardening," 8.

48. The house, at 120 East Bellevue Place, is now included in the National Register of Historic Places, and it is home to the Fortnightly, a private Chicago club. See Benjamin and Cohen, *Great Houses of Chicago*, 178.

49. Edward Renwick recollected Roche's arrival as "the Spring of 1882." Renwick Manuscript, 30. But a contemporary source, no doubt written in consultation with Roche himself, reported, "In May, 1872, he entered the office of W. L. B. Jenney, where he remained until 1881, when he made a partnership with Messrs. Holabird & Simonds, adding his name to the firm." Jenkins, "Review of the Work of Holabird & Roche," 1.

50. Renwick Manuscript, 29.

51. On Roche, see Bruegmann, *Architects and the City*, 12.

52. All quotations in this paragraph are from Renwick Manuscript, 24–27.

53. Renwick was made a partner in the firm in January 1896. Jenkins, "Review of the Work of Holabird & Roche," 1.

54. Graceland made an agreement with the railway in August 1882. Record Book, Graceland, 53 (31 August 1882).

55. Asher Carter had died in 1877, but Augustus Bauer, then in partnership with Henry W. Hill, still practiced in Chicago and presumably would have been available for the work. See Illinois Chapter of the American Institute of Architects, *In Memoriam Augustus Bauer*, 8.

56. At its December meeting, the board of managers resolved to erect "a sta-

tion and office building at Graceland . . . in accordance with the plans furnished by Holabird, Simonds and Roche." Record Book, Graceland, 55 (21 December 1882). See also "Graceland Cemetery—Cemetery Office and Station (entry 2)," in Bruegmann, *Holabird & Roche*, 2–3. Bruegmann reports that the building (now razed) originally included "Gentleman's and ladies' waiting rooms, toilet, and ticket office on the east; cemetery offices, to the west; two bedrooms . . . located above the west end of the building."

57. See "Furniture for Lathrop Office (entry 3)," in Bruegmann, *Holabird & Roche*, 3.

58. Renwick recollected that Simonds resigned in January 1883. Renwick Manuscript, 27. Simonds's obituary in the *Chicago Tribune* also lists 1883 as the year; see "O. C. Simonds, Park Designer, Taken by Death," *Chicago Tribune*, 22 November 1931, 16; see also Jenkins, "Review of the Work of Holabird & Roche," 1, which notes that the firm became Holabird & Roche that September. Simonds later wrote that the three men remained good friends; "Dean of the Cemetery Field," 21.

59. Later, along with new work at Graceland, Lathrop also commissioned Holabird & Roche for "alterations and additions" to his father's Elmhurst mansion, Huntington, and its gardener's cottage, along with a new barn. He also retained the firm to design a house for him in the "Ashland Subdivision" in Chicago. These projects likely date to the late 1880s. See Burnham Library–University of Illinois Project to Microfilm Architectural Documentation Records, 1950–52 (accession number 1952.1), reel 37, frames 94–124, Ryerson and Burnham Archives, Ryerson and Burnham Libraries, Art Institute of Chicago.

60. Simonds, "Graceland at Chicago," 12–13.

61. Remaining quotations in this paragraph are from Simonds, "Dean of the Cemetery Field," 20.

62. Simonds, "Graceland at Chicago," 12–13.

63. Simonds, *Landscape-Gardening*, v. Lathrop had died four years earlier, in 1916.

64. The essays are "A Plea for Landscape-Gardening" and "Parks and Landscape-Gardening."

65. Simonds, "Graceland at Chicago," 12.

66. All quotations in this paragraph are from Simonds, "Notes on Graceland," 8–9.

67. Simonds, "Graceland at Chicago," 12.

68. Ibid., 12 and 16.

69. Both quotations in this paragraph ibid., 12.

70. Lathrop, "A Plea for Landscape Gardening," 324.

71. All quotations in this paragraph are from Simonds, "Dean of the Cemetery Field," 21.

72. Ibid.

73. Ibid., 20.

74. "Ossian Cole Simonds" (*American Landscape Architect* obituary), 17.

75. Remaining quotations in this paragraph are from [Reich], "Graceland Cemetery," 2.

76. Reich also noted that Graceland's "walks, until 1880, and later, [were] only [made of] sand." Ibid.

77. Ibid.

78. Simonds, "Notes on Graceland," 9. Reich identified and explicitly attributed "the subdivision of Ridgeland into still larger lots" to Bryan Lathrop. See [Reich], "Graceland Cemetery," 2.

79. Renwick Manuscript, 26.
80. Of all the new burial sections created in Jenney's layout, only Lakeside is represented as subdivided in the 1884 *Map of Graceland Cemetery, Cook Co., Ill.*
81. The lower left corner of the map carries the legend: "Explanation. Re-sub-division of Part of Section G, M, N, P. Sections left blank are not yet subdivided into lots."
82. All quotations in the paragraph are from Simonds, "Notes on Graceland," 9.
83. Although Simonds did not identify it, this unnamed section is located at Graceland's southeast corner.
84. *Chicago Illustrated*, 50.
85. Simonds, "Graceland at Chicago," 16, and "Dean of the Cemetery Field," 19.
86. Simonds, "Dean of the Cemetery Field," 19. In "Graceland at Chicago," Simonds identified other sources for trees and shrubs. Among these were "Peterson's nursery"; the cemetery's original landscape gardener, Swain Nelson; Ellwanger & Barry of Rochester, N.Y.; and "Naperville Nurseries" (16).
87. Simonds, "Dean of the Cemetery Field," 20.
88. Simonds, "Notes on Graceland," 9.
89. Simonds, "Dean of the Cemetery Field," 20.
90. Simonds, "Notes on Graceland," 9.
91. Ibid.
92. Remaining quotations in this paragraph are from Simonds, "Landscaping Cemeteries," 187.
93. Creese, *Crowning of the American Landscape*, 209.
94. Ibid., 212.
95. "Graceland" (classified advertisement), *Chicago Tribune*, 28 September 1882, 1.

Epilogue
1. Andreas, *History of Cook County*, 721.
2. Bruegmann, *Architects and the City*, 33.
3. Ibid., 446.
4. On Olmsted and Richardson see, for instance, Mallgrave, *Modern Architectural Theory*, 162–63.
5. Simonds, *Landscape-Gardening*, 297.
6. Jenkins, "Review of the Work of Holabird & Roche," 34.
7. Ibid.
8. Ibid.
9. Simonds, "Notes on Graceland," 9.
10. Morrison, *Louis Sullivan*, 100.
11. Creese, *Crowning of the American Landscape*, 216.
12. Morrison, *Louis Sullivan*, 100.
13. Simonds, *Landscape-Gardening*, 305. The professional journal *Park and Cemetery* illustrated its obituary of Simonds with a photograph of the Ryerson monument, captioned "The beautiful cemetery lot pictures in Graceland are typical of Mr. Simonds' art." See "O. C. Simonds Passes On," *Park and Cemetery* 41, no. 10 (December 1931): 301.
14. "It Has a Crematory," *Chicago Tribune*, 25 November 1893, 3.
15. "How Chicago Looks upon the Custom of Cremation," *Chicago Tribune*, 17 April 1898, 29.
16. Simonds, "Dean of the Cemetery Field," 21.
17. "How Chicago Looks upon the Custom of Cremation," 29.
18. Bruegmann, *Architects and the City*, 451.

19. Simonds, *Landscape-Gardening*, 298.
20. A one-page promotional flyer issued by Graceland, "Medal Awarded to Graceland Cemetery by the Paris Exposition of 1900," was tipped in between pages 38 and 39 of the April 1915 issue of *Park and Cemetery and Landscape Gardening* (25, no. 2), accompanying the article "Spring Pictures in Graceland Cemetery." The text reads: "The Paris Exhibition of 1900 offered as prizes two medals for the best sets of twelve views in cemeteries of any country. The leading cemeteries of the world were represented by photographs of their grounds. Both prizes were awarded to American cemeteries, the gold medal to Spring Grove, Cincinnati, and the silver medal to Graceland, Chicago." The official reports of the exposition, however, suggest that this claim is inaccurate on several counts. There was no competition for cemetery or landscape design; the category the photographs were entered in was "Appliances and Methods of Horticulture and Arboriculture," a subclass of the general horticulture section (*Catalogue of Exhibitors*, 261–63). The informational circular issued by the director of that section of the U.S. exhibit, Charles Richards Dodge, specified twenty-four, not twelve, photographs; it explained that the main exhibit would consist of "large and representative" collections of "specimens of fresh fruits, vegetables, ornamental plants and seeds . . . canned, dried, and other manufactured products of fruits, nuts, and vegetables," but that an ancillary photograph display would allow "park commissioners, cemetery boards, florists, seedsmen, nurserymen, fruit growers, truck farmers, packers of canned or evaporated fruits and vegetables, etc., to have their establishments represented at the Exposition at a cost relatively small" (*Report of the Commissioner-General*, 2:231–32 [see also 3:352]; for the final report of the horticulture section see 3:446–56 and 6:355–75). The total number of medals awarded in the "Appliances and Methods" subclass was 303, including five *grands prix*, the highest category; no U.S. exhibit received this award (*Rapports du jury*, 13; for a list of all U.S. medal winners in the entire horticulture section see *Report of the Commissioner-General*, 6:357–58, and for European winners see the lists published in the *Revue Horticole* for that year). Spring Grove received one of 37 gold medals, and among the 107 silver medals three were awarded to U.S. cemeteries: Bellefontaine Cemetery in St. Louis (designed by Almerin Hotchkiss), Graceland, and Woodmere Cemetery in Detroit (designed, like Spring Grove, by Adolph Strauch). In its report the panel of judges of the subclass, which included the landscape architect Édouard André (1840–1911), praised the photographs submitted by U.S. cemeteries: "These cemeteries are absolutely different from our necropolises, where stone, marble, and bronze are piled one on another. . . . The inhabitants of the New World . . . have created immense parks where the roads are wide, where water winds around and over rocks, where the plantings are generous and . . . the lawns vast, . . . and within these huge parks that serve as a place for the living to walk, the dead rest in peace" (*Rapports du jury*, 12; see also *Report of the Commissioner-General*, 3:449 and 5:351–58). A subsequent mention of German cemeteries suggests that Germany may have been the only other country whose cemeteries were represented. I thank LALH editor and researcher Mary Bellino for locating and compiling the source material for this note. Correspondence between O. C. Simonds and the organizers of the horticulture exhibit is available at the National Archives in College Park, Md., in the Records of the Bureau of Plant Industry. I thank NARA archivist Joseph Schwarz for providing copies of this material.
21. "O. C. Simonds Passes On," 302.

22. Graceland Cemetery Company Board of Managers minute book, 98–101 (17 June 1903).

23. Ibid.

24. Burnham's friend, colleague, and biographer Charles Moore posthumously attributed this quotation to the architect; see Moore's *Daniel H. Burnham*, 2:147. On its origins see Hines, *Burnham of Chicago*, 401n8.

25. On Button see Debbie Lang's entry in Birnbaum and Foell, *Shaping the American Landscape*, 40–41. Simonds is known to have given new staff members "apprenticeship of service in the cemetery [Graceland] and on private landscape work on which he was engaged." J. Roy West (1880–1941), who was Simonds's partner after 1910, assisted him in training these younger men. See "J. Roy West: A Biographical Minute." George C. Cone (1868–1942) was another Simonds employee who probably worked on Graceland. Cone joined O. C. Simonds & Company in 1904. According to Cone's obituary, Simonds "employed Mr. Cone as his chief assistant in this phase of the work." See "George Carol Cone: A Biographical Minute."

26. Later, landscape architects of national repute, such as Warren H. Manning and Annette Hoyt Flanders, would also design monument surrounds in the cemetery. Manning prepared a garden surrounding for the plot of Cyrus McCormick Jr., and Flanders created "A Design for the Cemetery Plot of Mr. and Mrs. Philip D. Armour" (1936–39), which was likely implemented, although no trace of it remains now. The most extensive and accessible source on Flanders is Filzen, "Annette Hoyt Flanders."

27. Creese, *Crowning of the American Landscape*, 211.

28. Ibid., 212.

29. See "Odd Journey of a Tree," *Chicago Tribune*, 14 December 1890, 9; see also "Death of J. H. Lathrop," *Chicago Tribune*, 24 November 1889, 4. The tree has since been removed.

30. "M[iller], "An American Idea in Landscape Art."

31. Ibid., 194.

32. On Miller, Simonds, and the Prairie School see C. Vernon, "Introduction to the Reprint Edition," in the Library of American Landscape History edition of *The Prairie Spirit in Landscape Gardening*.

33. Medal of Honor certificate, dated 20 April 1925, Cemetery Office, Graceland.

34. The other was a residential project, "Locke Ledge," in Yorktown Heights, New York. Also that year, the name of the firm was changed from O. C. Simonds & Company to Simonds & West; see "O. C. Simonds Passes On," 302.

35. Newton, *Design on the Land*, 390–91.

36. Waugh, *Landscape Beautiful*, 194.

37. Pregill and Volkman, *Landscapes in History*, 506.

38. Eifler & Associates Architects, *Graceland Cemetery: A History* (Chicago, n.d.), 29.

39. Ibid., 31.

40. I thank longtime Graceland tour leader Diane Lanigan for this information.

41. Chris Jones, "'Graceland' Should Put Playwright on the Map," *Chicago Tribune*, 30 May 2009, 16.

42. Simonds, "Graceland at Chicago," 16.

43. Chappell, *Chicago's Urban Nature*, xix.

Bibliography

The main repositories of the historical records of Graceland Cemetery are the cemetery's main office and the Chicago History Museum. The museum holds two primary collections: the Graceland Cemetery Collection (1994.146), containing maps, plats, atlases, correspondence, documents, and drawings; and Architectural Photographs of Graceland Cemetery (1992.340). Citations for individual items in these and other collections, as well as newspaper articles and other ephemera, are given in the notes.

Abbott, Carl. "'Necessary Adjuncts to Growth': The Railroad Suburbs of Chicago." *Journal of the Illinois State Historical Society* 73, no. 2 (Summer 1980): 117–31.

Abrahamson, Sarah T. "Lake View: From Wilderness to Metropolis." *Illinois History* 49, no. 1 (December 1995): 22–23.

Alaimo, Marilyn K. National Register of Historic Places Nomination Form: Lake Forest Cemetery. Prepared for the City of Lake Forest. 2001. Available at www.illinoishistory.gov/ps/haargis.htm.

Allan, Mea. *William Robinson, 1838–1935: Father of the English Flower Garden.* London: Faber and Faber, 1982.

André, Edouard, *L'art des jardins: traité général de la composition des parcs et jardins.* 2 vols. Paris: G. Masson, 1879.

Andreas, A. T. *History of Chicago from the Earliest Period to the Present Time.* 3 vols. Chicago: A. T. Andreas, 1884–1886.

———. *History of Cook County, Illinois, from the Earliest Period to the Present Time.* Chicago: A. T. Andreas, 1884.

Andrews, Wayne. *Architecture, Ambition, and Americans: A Social History of American Architecture.* Rev. ed. New York: The Free Press, 1978.

"The Architect and the Landscape Gardener." *American Builder and Journal of Art* 2, no. 5 (May 1869): 100.

Arnold, Isaac N. *William B. Ogden and Early Days in Chicago.* Chicago: Fergus Printing Company, 1882.

"Asher Carter" (obituary). *American Architect and Building News* 2 (27 January 1877): 30–31.

Bachrach, Julia Sniderman. "Ossian Cole Simonds: Conservation Ethic in the Prairie Style." In Tishler, *Midwestern Landscape Architecture*, 80–98.

Bassman, Herbert J. *Riverside Then and Now.* 3rd ed. Riverside: Riverside Historical Commission, 1995.

Bates, Frederick H. *"Old Elmhurst": Being the Personal Recollections of a Native.* 1919; repr., Elmhurst: Elmhurst Historical Commission, 1973.

Bender, Thomas. "The 'Rural' Cemetery Movement: Urban Travail and the Appeal of Nature." *New England Quarterly* 47, no. 2 (June 1974): 196–211.

Benjamin, Susan, and Stuart Cohen. *Great Houses of Chicago, 1871–1921.* New York: Acanthus Press, 2008.

Berens, Helmut Alan. *Elmhurst: Prairie to Tree Town.* Elmhurst, Ill.: Elmhurst Historical Commission, 1968.

Beveridge, Charles E. "Frederick Law Olmsted's Theory of Landscape Design." *Nineteenth Century* 3, no. 2 (1977): 38–43.

Bigot, Mme. Charles [Mary Healy]. "Forgotten Chicago." In *Chicago Yesterdays: A Sheaf of Reminiscences,* edited by Caroline Kirkland, 143–61. Chicago: Daughaday and Company, 1919.

———. *Life of George P. A. Healy.* Privately printed, n.d.

Binns, Elizabeth. "Daniel Bryan: Poe's Poet of 'The Good Old Goldsmith School.'" *William and Mary College Quarterly Historical Magazine,* 2nd ser., 23, no. 4 (October 1943): 465–73.

Birnbaum, Charles A., and Stephanie S. Foell, eds. *Shaping the American Landscape: New Profiles from the Pioneers of American Landscape Design Project.* Charlottesville: University of Virginia Press, 2009.

Birnbaum, Charles A., and Robin Karson, eds. *Pioneers of American Landscape Design.* New York: McGraw-Hill, 2000.

Bisgrove, Richard. *William Robinson: The Wild Gardener.* London: Frances Lincoln, 2008.

Biographical Dictionary and Portrait Gallery of Representative Men of Chicago, Milwaukee and The World's Columbian Exposition. Chicago: American Biographical Publishing Company, 1892.

Bjorklund, Richard C., and Becky Haglund. *Pioneer Settler Conrad Sulzer.* Chicago: Ravenswood–Lake View Historical Association, 1986.

Blair, Erle O. "Ossian Cole Simonds." *Landscape Architecture* 22, no. 3 (April 1932): 234–35.

The Book of Chicagoans. Chicago: A. N. Marquis, 1905.

Brackett, G. B. "Horticulturists: Saunders, William." In *The Standard Cyclopedia of Horticulture,* edited by L. H. Bailey, 3:1594–95. New York: Macmillan, 1917.

"Brooklyn Park Improvements." *American Builder and Journal of Art* 2, no. 3 (March 1869): 71.

Brooks, H. Allen. *The Prairie School: Frank Lloyd Wright and His Midwest Contemporaries.* 1972; repr., New York: W. W. Norton, 1976.

Bruegmann, Robert. *The Architects and the City: Holabird & Roche of Chicago, 1880–1918.* Chicago: University of Chicago Press, 1997.

———. *Holabird & Roche, Holabird & Root: An Illustrated Catalogue of Works,* vol. 1: *1880–1911.* New York: Garland Publishing in cooperation with the Chicago Historical Society, 1991.

B[ryan, Thomas]. "Defence of the Prairies." *Horticulturist and Journal of Rural Art and Rural Taste* 14, no. 7 (July 1859): 341.

"Bryan Lathrop." In *Chicago Historical Society: Charter, Constitution, By-Laws, Membership List, Sixtieth Anniversary Year Book,* 173–75. Chicago: Chicago Historical Society, 1916.

Cannon, M. Hamlin. "The United States Christian Commission." *Mississippi Valley Historical Review* 38, no. 1 (June 1951): 61–80.

Catalogue of Exhibitors in the United States Sections of the International Universal Expo-

sition, Paris, 1900. Paris: Société Anonyme des Imprimeries Lemercier, 1900.

The Cemetery Handbook: A Manual of Useful Information on Cemetery Development and Management. Chicago: Allied Arts Publishing, 1921.

Chamberlin, Everett. *Chicago and Its Suburbs.* Chicago: T. A. Hungerford, 1874.

Chappell, Sally A. Kitt. *Chicago's Urban Nature: A Guide to the City's Architecture + Landscape.* Chicago: University of Chicago Press, 2007.

Charter, Rules and Regulations of the Rosehill Cemetery. Chicago: Office of the Cemetery, 1859.

Chicago Illustrated. Chicago: N. F. Hodson, 1883.

Clark, Stephen Bedell. *The Lake View Saga, 1837–2007.* Chicago: Lake View Citizen's Council, 2007.

Cleveland, H. W. S. "A Few Hints on the Arrangement of Cemeteries." *American Builder and Journal of Art* 2, no. 10 (October 1869): 187–88.

———. *Landscape Architecture as Applied to the Wants of the West.* 1873. Amherst: University of Massachusetts Press in association with Library of American Landscape History, 2002.

———. *The Public Grounds of Chicago: How to Give Them Character and Expression.* Chicago: Charles D. Lakey, 1869.

Condit, Carl W. *Chicago, 1910–29: Building, Planning, and Urban Technology.* Chicago: University of Chicago Press, 1973.

Conzen, Michael P. "Evolution of the Chicago Map Trade: An Introduction." *Chicago History* 13, no. 1 (Spring 1984): 4–11.

Cook, Frederick. *Bygone Days in Chicago: Recollections of the "Garden City" of the Sixties.* Chicago: A. C. McClurg, 1910.

Coventry, Kim, Daniel Meyer, and Arthur H. Miller. *Classic Country Estates of Lake Forest: Architecture and Landscape Design, 1856–1940.* New York: W. W. Norton, 2003.

Creese, Walter L. *The Crowning of the American Landscape: Eight Great Spaces and Their Buildings.* Princeton: Princeton University Press, 1985.

Cronon, William. *Nature's Metropolis: Chicago and the Great West.* New York: W. W. Norton, 1991.

Curl, James Steven. *A Celebration of Death: An Introduction to Some of the Buildings, Monuments, and Settings of Funerary Architecture in the Western European Tradition.* London: Constable, 1980.

Currey, J. Seymour. *Chicago: Its History and Its Builders.* Chicago: S. J. Clarke, 1912.

Dahlstrom, Neil, and Jeremy Dahlstrom. *The John Deere Story: A Biography of Plowmakers John and Charles Deere.* DeKalb: Northern Illinois University Press, 2005.

Darnall, Margaretta. "The American Cemetery as Picturesque Landscape: Bellefontaine Cemetery, St. Louis." *Winterthur Portfolio* 18, no. 4 (Winter 1983): 249–69.

Davenport, F. Garvin. "John Henry Rauch and Public Health in Illinois, 1877–1891." *Journal of the Illinois State Historical Society* 50 (1956): 277–93.

"Death of Augustus Bauer, Architect." *American Architect and Building News* 44, no. 962 (2 June 1894): 101.

De Mare, Marie. *G. P. A. Healy, American Artist: An Intimate Chronicle of the Nineteenth Century.* New York: David McKay, 1954.

Drury, John. *Old Chicago Houses.* Chicago: University of Chicago Press, 1941.

Dyson, Walter. *Howard University, the Capstone of Negro Education: A History, 1867–1940.* Washington, D.C.: The Graduate School, Howard University, 1941.

Ebert, Albert K. "Early History of the Drug Trade of Chicago, Compiled from the Records of the Chicago Veteran Druggist's Association." *Transactions of the Illinois State Historical Society* 8 (1903): 234–74.

Egbert, Donald D., and Paul E. Sprague. "In Search of John Edelmann" *AIA Journal* 45, no. 2 (February 1966): 35–41.

Eggener, Keith. "Building on Burial Ground." In *Cemeteries* (Library of Congress Visual Sourcebooks in Architecture, Design, and Engineering). New York: W. W. Norton, 2010.

Eilert, John W. "Illinois Business Incorporations, 1816–1869." *Business History Review* 37, no. 3 (Autumn 1963): 169–81.

Elmhurst Historical Commission. *Elmhurst: Scenes from Yesterday.* Elmhurst, Ill.: Elmhurst Historical Commission, 1986.

Eslinger, Ellen. "Gardening." Electronic Encyclopedia of Chicago, www.encyclopedia.chicagohistory.org/pages/498.html.

Freeman, Norman L. *Reports of Cases at Law and in Chancery, Argued and Determined in the Supreme Court of Illinois.* Vol. 44. Chicago: Callaghan and Company, 1886.

French, W. M. R. "Country Roads." *American Builder and Journal of Art* 3, no. 7 (July 1870): 152–53.

———. "Methods of Surveying Irregular Sub-divisions of Western Towns." *American Builder and Journal of Art* 4, no. 6 (June 1871): 415–17.

———. "Trees in Composition." *Fourth Volume of the American Park and Outdoor Art Association* (1900): 57–62.

Funigiello, Philip J. *Florence Lathrop Page: A Biography.* Charlottesville: University Press of Virginia, 1994.

Geiger, Barbara. "'Nature as the Great Teacher': The Life and Work of Landscape Designer O. C. Simonds." M.A. thesis, University of Wisconsin, Madison, 1997.

Gelbloom, Mara. "Ossian Simonds: Prairie Spirit in Landscape Gardening." *Prairie School Review* 12, no. 2 (Second Quarter 1975): 5–18.

Gemperle, Kathy. "The Raschers of Andersonville." *Edgewater Historical Society Scrapbook* 10, no. 3 (Fall 1999): n.p.

"George Carol Cone: A Biographical Minute." *Landscape Architecture* 32, no. 4 (July 1942): 164–65.

Geraniotis, Roula Mouroudellis. "An Early German Contribution to Chicago's Modernism." In *Chicago Architecture, 1872–1922: Birth of a Metropolis*, edited by John Zukowsky, 91–105. Munich: Prestel-Verlag in association with the Art Institute of Chicago, 1987.

———. "German Architectural Theory and Practice in Chicago, 1850–1900." *Winterthur Portfolio* 21, no. 4 (Winter 1986): 293–306.

Gerdts, William H. "'Chicago Is Rushing Past Everything': The Rise of American Art Journalism in the Midwest, from the Development of the Railroad to the Chicago Fire." *American Art Journal* 27, nos. 1–2 (1995–1996): 38–83.

Gilpin, William. *Three Essays: On Picturesque Beauty; On Picturesque Travel; and On Sketching Landscape; to which is added a poem on landscape painting.* London: R. Blamire, 1792.

Gilpin, William Sawrey. *Practical Hints upon Landscape Gardening: With Some Remarks on Domestic Architecture as Connected with Scenery.* London: T. Cadell, 1832.

Goode, Patrick, and Michael Lancaster, eds. *The Oxford Companion to Gardens.* Oxford: Oxford University Press, 1986.

Graceland Cemetery. Chicago: Photographic Print Co., 1904.

[Graceland Cemetery Company]. *Catalogue of the Graceland Cemetery Lot Owners to April 1870.* Chicago: Office of the Graceland Cemetery, 1870.

[———]. *Charter of the Graceland Cemetery (Approved Feb. 22, 1861) with the Rules and Regulations of the Company.* Chicago: James Barnet, 1861.

[————]. *The Proposition of the Graceland Cemetery Company and the Report Thereon of the Committee of the Citizens to the Town of Lake View.* Chicago: J. M. W. Jones, 1879.

Graceland Cemetery Company and Crematorium. *Historical Sketch of Graceland Cemetery, Chicago.* Chicago, n.d. [1960s?].

"Graceland: The Men Who Made Chicago." *Abitare* 256 (July–August 1987): 114–17.

Greeley, Samuel S. "The Laying Out of Western Suburban Villages." *American Builder and Journal of Art* 4, no. 4 (April 1871): 370–72.

Green, Constance McLaughlin. *Washington: A History of the Capital, 1800–1950.* Princeton: Princeton University Press, 1962.

Greiff, Constance M. *John Notman, Architect, 1810–1865.* Philadelphia: The Athenaeum of Philadelphia, 1979.

Grese, Robert E. "Introduction to the Reprint Edition." In Simonds, *Landscape-Gardening,* ix–lvi.

————. *Jens Jensen: Maker of Natural Parks and Gardens.* Baltimore: Johns Hopkins University Press, 1992.

————. "The Prairie Gardens of O. C. Simonds and Jens Jensen." In *Regional Garden Design in the United States,* edited by Therese O'Malley and Marc Treib, 99–124. Washington, D.C.: Dumbarton Oaks Research Library and Collection, 1995.

G[ridley], A. D. "Rural Cemeteries." *Horticulturist and Journal of Rural Art and Rural Taste* 5, no. 6 (June 1855): 278–82.

Griffiths, Mark. *The Lotus Quest.* London: Chatto & Windus, 2009.

Groce, George C., and David H. Wallace. *The New-York Historical Society's Dictionary of Artists in America, 1564–1860.* New Haven: Yale University Press, 1957.

Haglund, Karl. "Rural Tastes, Rectangular Ideas, and the Skirmishes of H. W. S. Cleveland." *Landscape Architecture* 66, no. 1 (January 1976): 67–70, 78.

Healy, George P. A. *Reminiscences of a Portrait Painter.* Chicago: A. C. McClurg, 1894.

Heathcote, Edwin. *Monument Builders: Modern Architecture and Death.* London: Academy Editions, 1999.

Hines, Thomas S. *Burnham of Chicago: Architect and Planner.* 2nd ed. Chicago: University of Chicago Press, 2008.

Historic Rock Island County: History of the Settlement of Rock Island County from the Earliest Known Period to the Present Time. Rock Island, Ill.: Kramer & Company, 1908.

Holland, Robert A. *Chicago in Maps, 1612–2002.* New York: Rizzoli, 2002.

Holland's Moline Directory for the Centennial Year, 1876, Containing a Historical Sketch of the City. Chicago: Holland Publishing, 1876.

Holland's Rock Island and Moline Directory for the Years 1882–85, Containing a Historical Sketch of the City. Chicago: Holland Publishing, 1882.

"A House in the West." *Horticulturist and Journal of Rural Art and Rural Taste* 15, no. 1 (January 1860): 12–13.

Hubbard, Henry Vincent, and Theodora Kimball. *An Introduction to the Study of Landscape Design.* New York: Macmillan, 1917.

Hubbard, Theodora Kimball. "H. W. S. Cleveland: An American Pioneer in Landscape Architecture and City Planning." *Landscape Architecture* 20, no. 2 (January 1930): 92–111.

Hudson, John C. "Topography." Electronic Encyclopedia of Chicago, www.encyclopedia.chicagohistory.org/pages/1260.html.

Hunt, John Dixon. *The Picturesque Garden in Europe.* London: Thames and Hudson, 2002.

Illinois Chapter of the American Institute of Architects. *In Memoriam Augustus Bauer: Died February 8, 1894*. Chicago, 1894.

"J. Roy West: A Biographical Minute." *Landscape Architecture* 32, no. 3 (April 1942): 120–21.

Jackson, John Brinckerhoff. *American Space: The Centennial Years, 1865–1876*. New York: W. W. Norton, 1972.

Jackson, Kenneth T., and Camilo Jose Vergara. *Silent Cities: The Evolution of the American Cemetery*. New York: Princeton Architectural Press, 1989.

Jenkins, Charles E. "A Review of the Work of Holabird & Roche." *Architectural Reviewer* 3 (June 1897): 1–41.

[Jenney, William Le Baron]. "Autobiography of William Le Baron Jenney, Architect." *Western Architect* 10, no. 6 (June 1907): 59–66.

Johnson, Allen, and Dumas Malone, eds. *Dictionary of American Biography*. New York: Charles Scribner's Sons, 1958.

Journal of the House of Representatives of the Twenty-Fourth General Assembly of the State of Illinois, at their regular session, begun and held at Springfield, January 2, 1865. Springfield, Ill.: Baker & Phillips, 1865.

Journal of the Senate of the Twenty-Fourth General Assembly of the State of Illinois, at their regular session, begun and held at Springfield, January 2, 1865. Springfield, Ill.: Baker & Phillips, 1865.

Karamanski, Theodore J. "Civil War." Electronic Encyclopedia of Chicago, www.encyclopedia.chicagohistory.org/pages/2379.html.

———. *Rally 'Round the Flag: Chicago and the Civil War*. Chicago: Nelson-Hall, 1993.

Keating, Ann Durkin. *Building Chicago: Suburban Developers and the Creation of a Divided Metropolis*. Columbus: Ohio State University Press, 1988.

Kern, G. M. *Practical Landscape Gardening, with reference to the Improvement of Rural Residences, giving the General Principles of the Art; with full directions for Planting Shade Trees, Shrubbery and Flowers, and Laying Out Grounds*. Cincinnati: Moore, Wilstach, Keys, 1855.

Kiefer, Charles D., Rolf Achilles, and Neal A. Vogel. National Register of Historic Places Nomination Form: Graceland Cemetery & Crematorium. Prepared for Legacy Historic Preservation Planning Services. 2000. Available at www.illinoishistory.gov/ps/haargis.htm.

Kirkland, Moses. *History of Chicago, Illinois*. Vol. 2. Chicago: Munsell, 1895.

Knoblauch, Marion, ed. *Du Page County: A Descriptive and Historical Guide, 1831–1939*. Elmhurst, Ill.: Irvin A. Ruby, 1948.

Kowsky, Francis R. *Country, Park & City: The Architecture and Life of Calvert Vaux*. New York: Oxford University Press, 1998.

Larson, Erik. *The Devil in the White City: Murder, Magic, and Madness at the Fair That Changed America*. New York: Vintage Books, 2004.

Lathrop, Bryan. "Parks and Landscape-Gardening." In *Park Commissioners' Session of the Sixth Annual Meeting, Boston, Aug. 5, 6, 7, 1902*, 7–10. Rochester: American Park and Outdoor Art Association, 1903. Reprinted in Simonds, *Landscape-Gardening*, 325–31.

Lathrop, Bryan. "A Plea for Landscape-Gardening." In Simonds, *Landscape-Gardening*, 323–25.

Leland, Ernest Stevens, and Donald W. Smith. "Ossian Cole Simonds: Master of Landscape Architecture." In *The Pioneers of Cemetery Administration in America: A Collection of Biographical Essays*. N.p.: "Privately printed and dedicated to the Association of American Cemetery Superintendents," 1941.

Lewis, Michael J. "The First Design for Fairmount Park." *Pennsylvania Magazine of History and Biography* 130, no. 3 (July 2006): 283–97.

Linden, Blanche M. G. *Silent City on a Hill: Picturesque Landscapes of Memory and Boston's Mount Auburn Cemetery.* 2nd ed. Amherst: University of Massachusetts Press in association with Library of American Landscape History, 2007.

Logan, Rayford Whittingham. *Howard University: The First Hundred Years, 1867–1967.* New York: New York University Press, 1969.

Loring, Sanford E., and W. L. B. Jenney. *Principles and Practice of Architecture: Comprising forty-six folio plates of plans, elevations and details of Churches, Dwellings and Stores constructed by the authors. Also, an explanation and illustrations of the French System of Apartment Houses, and Dwellings for the Laboring Classes, together with copious text.* Chicago: Cobb, Pritchard, 1869.

Loudon, J. C. *On the Laying Out, Planting, and Managing of Cemeteries and on the Improvement of Churchyards.* 1843. Redhill, Surrey: Ivelet Books, 1981.

Major, Judith K. *To Live in the New World: A. J. Downing and American Landscape Gardening.* Cambridge: MIT Press, 1997.

Mallgrave, Harry Francis. *Modern Architectural Theory: A Historical Survey, 1673–1968.* Cambridge: Cambridge University Press, 2005.

Maloney, Cathy Jean. *Chicago Gardens: The Early History.* Chicago: University of Chicago Press, 2008.

Marshall, Helen E. *Grandest of Enterprises: Illinois State University, 1857–1957.* Normal: Illinois State Normal University, 1956.

McCarthy, Kathleen D. *Noblesse Oblige: Charity and Cultural Philanthropy in Chicago, 1849–1929.* Chicago: University of Chicago Press, 1982.

Meehan, Joseph. "National Cemetery, Gettysburg." *Park and Cemetery* 7, no. 5 (July 1897): 104–6.

Memorial Book of the Old Tippecanoe Club, Chicago. Chicago: Peerless Printing Co., 1888.

Menocal, Narciso G. *Architecture as Nature: The Transcendentalist Idea of Louis Sullivan.* Madison: University of Wisconsin Press, 1981.

———. "The Iconography of Architecture: Sullivan's View." In *Louis Sullivan: The Poetry of Architecture,* edited by Robert Twombly and Narciso G. Menocal, 73–160. New York: W. W. Norton, 2000.

Miller, Donald L. *City of the Century: The Epic of Chicago and the Making of America.* 1996. New York: Simon & Schuster, 2003.

Miller, Naomi. *Heavenly Caves: Reflections on the Garden Grotto.* New York: George Braziller, 1982.

M[iller], W[ilhelm]. "An American Idea in Landscape Art." *Country Life in America* 4, no. 5 (September 1903): 349–50.

———. *The Prairie Spirit in Landscape Gardening.* 1915. Amherst: University of Massachusetts Press in association with Library of American Landscape History, 2002.

Moore, Charles. *Daniel H. Burnham, Architect, Planner of Cities.* 2 vols. Boston: Houghton Mifflin, 1921.

Morrison, Hugh. *Louis Sullivan: Prophet of Modern Architecture.* 1935. New York: W. W. Norton, 1998.

Nadenicek, Daniel J. "Nature in the City: Horace Cleveland's Aesthetic." *Landscape and Urban Planning* 26 (1993): 5–15.

———. "Sleepy Hollow Cemetery: Philosophy Made Substance." *Emerson Society Papers* 5, no. 1 (Spring 1994): 1–2, 8.

———. "Sleepy Hollow Cemetery: Transcendental Garden and Community Park." *Journal of the New England Garden History Society* 3 (Fall 1993): 8–15.

Nadenicek, Daniel J., and Lance M. Neckar. "Introduction to the Reprint Edition." In Cleveland, *Landscape Architecture as Applied to the Wants of the West,* xi–lxxii.

The National Cyclopaedia of American Biography. New York: James T. White and Company, 1893.

Newton, Norman T. *Design on the Land: The Development of Landscape Architecture*. Cambridge: Harvard University Press, 1971.

North, Edward. "The Proper Expression of a Rural Cemetery." *Horticulturist and Journal of Rural Art and Rural Taste* 7, no. 6 (June 1857): 253–56.

"O. C. Simonds Passes On." *Park and Cemetery* 41, no. 10 (December 1931): 301–2.

Olmsted, Vaux & Co. *Preliminary Report upon the Proposed Suburban Village at Riverside, Near Chicago* (1868). In *The Papers of Frederick Law Olmsted*, vol. 6, *The Years of Olmsted, Vaux & Company, 1865–1874*, edited by David Schuyler and Jane Turner Censer, 273–90. Baltimore: Johns Hopkins University Press, 1992.

"Ossian C. Simonds." *American Cemetery* 3, no. 12 (December 1931): 14–15.

"Ossian Cole Simonds." *American Landscape Architect* 5, no. 6 (December 1931): 17.

Pattison, William D. "The Cemeteries of Chicago: A Phase of Land Utilization." *Annals of the Association of American Geographers* 45, no. 3 (September 1955): 245–57.

———. "Land for the Dead of Chicago." Ph.D. diss., University of Chicago, 1952.

"Personal." *American Builder and Journal of Art* 2, no. 5 (May 1869): 116.

Peters, Harry T. *America on Stone: The Other Printmakers to the American People*. New York: Doubleday, Doran, 1931.

Picturesque Chicago. Chicago: Chicago Engraving Company, 1882.

Pond, Irving K. *The Autobiography of Irving K. Pond: The Sons of Mary and Elihu*. Edited by David Swan and Terry Tatum. Oak Park, Ill.: Hyoogen Press, 2009.

Pregill, Philip, and Nancy Volkman. *Landscapes in History: Design and Planning in the Eastern and Western Traditions*. New York: John Wiley & Sons, 1999.

Pressman, Lenore. *Graceland Cemetery: A Historical and Artistic Guide*. Chicago: Illinois Arts Council, 1976.

Private Laws of the State of Illinois, passed by the Twenty-Fourth General Assembly, convened January 2, 1865. Vol. 1. Springfield: Baker & Phillips, 1865.

Proceedings of the Board of Education of the State of Illinois: Regular Annual Meeting Held at Normal, June 6, 1917. Springfield: State of Illinois, 1917.

Proceedings of the Board of Education of the State of Illinois: Regular Meeting Held at Normal, December 16th, 1868. Peoria: N. C. Nason, Office of the Illinois Teacher, 1869.

Rainey, Reuben M. "A Prairie Metamorphosis: William Le Baron Jenney's Vision of Chicago's Central Park." *Threshold: Journal of the Chicago School of Architecture, The University of Illinois at Chicago* (New York: Rizzoli, 1991), 39–60.

———. "William Le Baron Jenney and Chicago's West Parks: From Prairies to Pleasure-Grounds." In Tishler, *Midwestern Landscape Architecture*, 57–79.

Randall, Frank A. *History of the Development of Building Construction in Chicago*. Rev. ed. Urbana: University of Illinois Press, 1999.

Ranney, Victoria Post. "Frederick Law Olmsted: Designing for Democracy in the Midwest." In Tishler, *Midwestern Landscape Architecture*, 41–56.

Rapports du jury international de l'Exposition Universelle de 1900 à Paris: Groupe VIII – Horticulture. Paris: Imprimerie Nationale, 1902.

Rauch, John H. *Intramural Interments in Populous Cities and Their Influence upon Health and Epidemics*. Chicago: Tribune Company, 1866.

———. *Public Parks: Their Effects upon the Moral, Physical and Sanitary Condition of the Inhabitants of Large Cities, with special reference to the City of Chicago*. Chicago: S. C. Griggs, 1869.

"Dr. [John H.] Rauch." "Public Parks." *American Builder and Journal of Art* 2, no. 4 (April 1869): 83–84.

Rayne, Mrs. M. L. *Chicago and One Hundred Miles Around: Being a Complete Handbook and Guide to the Public and Private Institutions, Churches, Schools, Libraries, Asylums, Railroad Offices, etc., etc., of the Garden City.* Chicago: R. Edwards, Printer, 1865.

Report of the Commissioner-General for the United States to the International Universal Exposition, Paris, 1900. 6 vols. Washington, D.C.: Government Printing Office, 1901.

Reps, John W. *Views and Viewmakers of Urban America: Lithographs of Towns and Cities in the United States and Canada, Notes on the Artists and Publishers, and a Union Catalogue of Their Work, 1825–1925.* Columbia: University of Missouri Press, 1984.

Revised Ordinances of the Town of Lake View. Lake View, Ill.: Town of Lake View, 1879.

Revue Horticole: 72° Anée – 1900. Paris: Librarie Agricole de la Maison Rustique, 1900.

Reynolds, William C. *In the Supreme Court of Illinois, Northern Grand Division, September Term, A.D. 1877, The People ex rel. Louis C. Huck, Country Collector, &c., vs. Graceland Cemetery Co. Amended Abstract and Argument for Appellee.* Chicago: Beach, Barnard & Co., Legal Printers, 1877.

———. *The Limit of the Police Power in the Control of Corporations. A Statement of the Condition, Property, and Franchises, of the Graceland Cemetery Co., together with a Review of the Legislation Concerning Cemeteries in Lake View.* Chicago: Chicago Legal News Company, 1872.

Ristow, Walter W. *American Maps and Mapmakers: Commercial Cartography in the Nineteenth Century.* Detroit: Wayne State University Press, 1985.

Riverside Improvement Company. *Riverside in 1871, with a Description of Its Improvements Together with Some Engravings of Views and Buildings.* Chicago: D. & C. H. Blakely, 1871.

"Riverside Park, Chicago." *Horticulturist and Journal of Rural Art and Rural Taste* 25 (November 1870): 325–27.

Robinson, William. "Garden Cemeteries." *The Garden* 10 (19 August 1876): 186–88.

———. *God's Acre Beautiful; or, The Cemeteries of the Future.* 3rd ed. London: John Murray, 1883.

———. *The Parks, Promenades, & Gardens of Paris: Considered in relation to the wants of cities and of public and private gardens; being notes on a study of parks and gardens.* London: Macmillan, 1878.

Russell, Don. *Elmhurst: Trails From Yesterday.* Elmhurst, Ill.: City of Elmhurst, 1977.

Russo, Edward J., and Curtis R. Mann. *Oak Ridge Cemetery.* Images of America series. Charleston, S.C.: Arcadia Publishing, 2009.

Rybczynski, Witold. *A Clearing in the Distance: Frederick Law Olmsted and America in the 19th Century.* New York: Scribner, 1999.

Salway, Wm. "Cemetery of Spring Grove, Cincinnatti." *Park and Cemetery* 5, no. 1 (March 1895): 4–7.

Saunders, William. "Designs for Improving Country Residences." *Horticulturist and Journal of Rural Art and Rural Taste* 5, no. 9 (September 1855): 403–5.

———. "Plan of Hunting Park, between the Built Part of Philadelphia and Germantown." *Horticulturist and Journal of Rural Art and Rural Taste* 8 (October 1858): 460–64.

———. "Planting of Roads and Avenues." *Park and Cemetery* 7, no. 2 (April 1897): 30–31.

Schmitt, Peter J. *Back to Nature: The Arcadian Myth in Urban America*. Baltimore: Johns Hopkins University Press, 1990.

Schuyler, David. *The New Urban Landscape: The Redefinition of City Form in Nineteenth-Century America*. Baltimore: Johns Hopkins University Press, 1986.

Schuyler, David, and Jane Turner Censer, eds. *The Papers of Frederick Law Olmsted*, vol. 6: *The Years of Olmsted, Vaux & Company, 1865–1874*. Baltimore: Johns Hopkins University Press, 1992.

Sclair, Helen. "Cemeteries." Electronic Encyclopedia of Chicago, www.encyclopedia.chicagohistory.org/pages/223.html.

———. "The Story of Graceland: A Prairie Landscape." Typescript, 1995. Copy held at the Graceland Cemetery office.

Scott, Frank J. *The Art of Beautifying Suburban Home Grounds of Small Extent*. New York: D. Appleton, 1870.

Scott, Valerie, ed. *Tooley's Dictionary of Mapmakers*. Riverside, Conn.: Early World Press, 2004.

Scott, William H. "Village Cemeteries." *Horticulturist and Journal of Rural Art and Rural Taste* 5, no. 4 (April 1855): 174–77.

Seligman, Amanda. "Uptown." Electronic Encyclopedia of Chicago, www.encyclopedia.chicagohistory.org/pages/1293.html.

Simo, Melanie Louise. *Loudon and the Landscape: From Country Seat to Metropolis, 1783–1843*. New Haven: Yale University Press, 1988.

Simon, Andreas, ed. *Chicago, the Garden City: Its Magnificent Parks, Boulevards, and Cemeteries*. Chicago: Franz Gindele, 1893.

Simonds, Ossian Cole. "The Dean of the Cemetery Field." *American Cemetery* (September 1930): 19–20, 37.

———. "A Few Words from a Landscape Gardener." *Park and Cemetery* 5, no. 5 (July 1895): 74–76.

———. "Graceland at Chicago." *American Landscape Architect* 6, no. 1 (January 1932): 12–16.

———. "Ground Coverings in the Cemetery." *Park and Cemetery* 24, no. 8 (October 1917): 263–65.

———. "How to Develop Beauty and Seclusion in Cemetery Design." *Park and Cemetery* 41, no. 5 (July 1931): 144.

———. "The Landscape-Gardener and His Work." *Park and Cemetery* 7, no. 1 (March 1897): 3–5.

———. *Landscape-Gardening*. 1920. Amherst: University of Massachusetts Press in association with Library of American Landscape History, 2000.

———. "Landscaping Cemeteries." *Monument and Cemetery Review* 16, no. 7 (March 1931): 185–88.

———. "Nature as the Great Teacher in Landscape Gardening." *Landscape Architecture* 22, no. 2 (January 1932): 100–108.

———. "Nature in the Cemetery." *Park and Cemetery* 35, no. 8 (October 1925): 208–9.

———. "Notes on Graceland." *American Landscape Architect* 2, no. 5 (May 1930): 8–9.

———. "The Planning and Administration of a Landscape Cemetery." *Country Life in America* 4, no. 5 (September 1903): 350.

———. "Progress of Rural Cemeteries." *Park and Cemetery* 33, no. 9 (November 1923): 234–36.

———. "The Use of Shrubs in Cemeteries." *Park and Cemetery* 10, no. 8 (October 1900): 176–79.

Sinkevitch, Alice, ed. *The AIA Guide to Chicago*. New York: Harcourt Brace, 1993.

Smith, J[ohn] J[ay]. *Designs for Monuments and Mural Tablets: Adapted to Rural Cemeteries, Church Yards, Churches and Chapels. With a Preliminary Essay on the*

Laying Out, Planting and Managing of Cemeteries and on the Improvement of Church Yards on the Basis of Loudon's Work. New York: Bartlett & Welford, 1846.

[————]. "Editor's Table: To Contributors and Exchanges, &c., &c." *The Horticulturist and Journal of Rural Art and Rural Taste* 14, no. 4 (April 1859): 185–91.

[————]. "Editor's Table: Rural Cemeteries and Public Parks." *Horticulturist and Journal of Rural Art and Rural Taste* 14, no. 10 (October 1859): 568.

————. "Our Chicago Correspondence." *Horticulturist and Journal of Rural Art and Rural Taste* 14, no. 5 (May 1859): 244–46.

————. *Recollections of John Jay Smith.* Philadelphia: J. B. Lippincott, 1892.

————. "Rural Cemeteries." *Horticulturist and Journal of Rural Art and Rural Taste* 6, no. 8 (August 1856): 345–47.

————. "Rural Cemeteries." *Horticulturist and Journal of Rural Art and Rural Taste* 8, no. 7 (July 1858): 297–300.

————. "Rural Cemeteries, No. 2. Planting, &c." *Horticulturist and Journal of Rural Art and Rural Taste* 6, no. 9 (September 1856): 393–96.

————. "Rural Cemeteries, No. 3. Conclusion." *Horticulturist and Journal of Rural Art and Rural Taste* 6, no. 10 (October 1856): 441–44.

Sniderman, Julia, Bart Ryckbosch, and Laura Taylor. National Register of Historic Places Nomination Form: Lincoln Park. Prepared for the Chicago Park District. 1994. Available at www.illinoishistory.gov/ps/haargis.htm.

Spencer, W. S., and David D. Griswold. *W. S. Spencer's Chicago Business Directory, for 1864–65. Containing the Names of All Firms and Individuals Doing Business in the City, Arranged in Alphabetical Order, and Classified.* Chicago: J. W. Dean's Book and Job Printing Establishment, 1864.

Spring Grove Cemetery: Its History and Improvements with Observations on Ancient and Modern Places of Sepulchre. Cincinnati: Robert Clarke, 1869.

Sullivan, Louis H. *The Autobiography of an Idea.* 1924; repr., New York: Dover, 1956.

Tatum, George B., and Elisabeth Blair MacDougall, eds. *Prophet with Honor: The Career of Andrew Jackson Downing, 1815–1852.* Washington, D.C.: Dumbarton Oaks Research Library and Collection and the Athenaeum of Philadelphia, 1989.

Taylor, Patrick, ed. *The Oxford Companion to the Garden.* Oxford: Oxford University Press, 2006.

Taylor, William A. "Awards at the Paris Exposition." *American Gardening* 21, no. 300 (September 22, 1900): 628.

Tishler, William H., ed. *Midwestern Landscape Architecture.* Urbana: University of Illinois Press in association with Library of American Landscape History, 2000.

Tolzmann, Don Heinrich, ed. *Spring Grove and Its Creator: H. A. Rattermann's Biography of Adolph Strauch.* Cincinnatti: The Ohio Book Store, 1988.

Turak, Theodore. "The École Centrale and Modern Architecture: The Education of William Le Baron Jenney." *Journal of the Society of Architectural Historians* 29, no. 1 (March 1970): 40–47.

————. "Jenney's Lesser Works: Prelude to the Prairie Style?" *Prairie School Review* 7, no. 3 (Third Quarter 1970): 5–20.

————. "Riverside: Roots in France." *Inland Architect* 25, no. 9 (November/December 1981): 12–19.

————. *William Le Baron Jenney: A Pioneer of Modern Architecture.* Ann Arbor: UMI Research Press, 1986.

————. "William Le Baron Jenney: Pioneer of Chicago's West Parks." *Inland Architect* 25, no. 2 (March 1981): 39–45.

————. "William Le Baron Jenney: Teacher." *Threshold: Journal of the University of Illinois at Chicago School of Architecture* 5/6 (Fall 1991): 61–81.

Twombly, Robert. *Louis Sullivan: His Life and Work.* New York: Viking, 1986.

University of Michigan. *Calendar of the University of Michigan, 1874/75.* Ann Arbor: University of Michigan, 1874.

———. *General Register, 1876–1877.* Ann Arbor: University of Michigan, 1876.

Van Zanten, David. "The Nineteenth Century: The Projecting of Chicago as a Commercial City and the Rationalization of Design and Construction." In *Chicago and New York: Architectural Interactions,* 30–48. Chicago: Art Institute of Chicago, 1984.

———. "Sullivan to 1890." In *Louis Sullivan: The Function of Ornament,* edited by Wim de Wit, 13–63. New York: W. W. Norton, 1986.

Vernon, Christopher. "Introduction to the Reprint Edition." In Miller, *The Prairie Spirit in Landscape Gardening,* ix–xxx.

Vernon, Noël Dorsey. "Adolph Strauch: Cincinnati and the Legacy of Spring Grove Cemetery." In Tishler, *Midwestern Landscape Architecture,* 5–24.

Vinci, John. "Graceland: The Nineteenth-Century Garden Cemetery." *Chicago History* 6, no. 2 (Summer 1977): 86–98.

Voss, Frederick. "Webster Replying to Hayne: George Healy and the Economics of History Painting." *American Art* 15, no. 3 (Autumn 2001): 34–53.

[Walker, Edwin S.] *Oak Ridge Cemetery: Its History and Improvements, Rules and Regulations.* Springfield, Ill.: H. W. Rokker, 1879.

Waller, James B. *Right of Eminent Domain and Police Power of the State. Trustees of Lake View Township, against Graceland Cemetery Company. Argument by James B. Waller.* Chicago: Board of Trustees, Town of Lake View, and Jameson & Morse, Printers, 1871.

Ware, Benjamin P. "Memorial of William Saunders." *Transactions of the Massachusetts Horticultural Society for the Year 1901, Part 1,* 11–12. Boston: For the Society, 1902.

Waugh, Frank A. *The Landscape Beautiful: A Study in the Utility of the Natural Landscape, Its Relation to Human Life and Happiness, with the Application of These Principles in Landscape Gardening, and in Art in General.* New York: Orange Judd, 1910.

Who Was Who in America. Vol. 1. Chicago: A. N. Marquis, 1943.

Who Was Who in America: Historical Volume, 1607–1896. Chicago: A. N. Marquis, 1963.

Wight, Peter B. "A Portrait Gallery of Chicago Architects: IV. Asher Carter." *Western Architect* 34, no. 1 (January 1925): 10–13.

Wills, Gary. *Lincoln at Gettysburg: The Words That Remade America.* New York: Simon & Schuster, 1992.

Winkler, Franz [pseud. Montgomery Schuyler]. "Some Chicago Buildings Represented by the Work of Holabird & Roche." *Architectural Record* 31, no. 4 (April 1912): 313–88.

Worpole, Ken. *Last Landscapes: The Architecture of the Cemetery in the West.* London: Reaktion Books, 2003.

Wunsch, Aaron. "Emporia of Eternity: 'Rural' Cemeteries and Urban Goods in Antebellum Philadelphia." *Nineteenth Century* 28, no. 2 (Fall 2008): 14–23.

Yalom, Marilyn. *The American Resting Place: 400 Years of History through Our Cemeteries and Burial Grounds.* Boston: Houghton Mifflin, 2008.

Yearbook of the United States Department of Agriculture, 1900. Washington: Government Printing Office, 1901.

Young, William, ed. *A Dictionary of American Artists, Sculptors and Engravers.* Cambridge, Mass.: William Young, 1968.

Index